# SECRET WARFARE

# SECRET WARFARE

## The Arms and Techniques of the Resistance

Pierre Lorain

*English Adaptation by David Kahn*

Orbis Publishing · London

First published in Great Britain by Orbis Publishing Limited, London 1984



Printed in Great Britain by R. J. Acford.

ISBN 0-85613-586-0

*This work is dedicated to all those who chose to fight in the night to save Truth and Reason.*

*Let this fraternity of men-at-arms find light beyond the terrible years, in time of peace.*

# Contents

# Editor's Foreword

When a friend brought me *Armement Clandestin*, the French original of this book, I was immediately struck, as I am sure readers will be, by the clear and elegant line drawings of guns and airplanes. But I am no expert on these things, and what suggested to me that this was a book of unusual quality and excellence was the part on ciphers, about which I do know something.

All too often the section on cryptology in a book on spying is disappointing. Errors abound and the phraseology betrays that the author has not really grasped the subject: had he done so, he would have expressed things differently.

The case with *Armement Clandestin* was entirely the reverse. Not only did the text show that Pierre Lorain knew the specifics of the cryptosystems with a remarkable precision, but it also demonstrated that he had such a comprehension of the overall field and of its development that he was able to generalize and bring some fresh observations to his text. Beyond this tone, which was so refreshing, the book provided a great many new details about the cryptosystems used between the French Resistance networks and Great Britain's Special Operations Executive, the agency that aided these movements and received intelligence from them.

All this impressed me, and I reasoned that if Lorain had done so careful a job with this relatively recondite area, he must have done so as well with other subjects. (This has proved true. For example, many gun experts believe that the Liberator handgun, which American factories stamped out by the hundreds of thousands for underground use, was a major weapon of the French Resistance. Lorain investigated the matter, collaborating with an American researcher. They satisfied themselves that no Liberator ever reached the Resistance, probably owing to the gun's poor quality. He consequently kept it out of his book.) Moreover, I knew of no other publication like *Armement Clandestin*, which details the nuts-and-bolts equipment and procedures of agents in the field. I thought it would make a valuable contribution to the history of underground movements, guerrilla warfare, armaments, and World War II. So I recommended to the Macmillan Publishing Company that it issue an English translation, and it agreed.

Pierre Lorain is an architect living in Paris. He is an expert on antique weapons and writes frequently on the subject for French magazines, often illustrating them with drawings, for which his architect's training has proved useful. He is an amateur radio operator, which enabled him to understand the World War II sets and their capabilities. He wrote and illustrated *Armement Clandestin,* which was published privately in France in 1972, as a labor of love and as a memorial to the men and women who fought underground in World War II for the freedom of France.

As the English-language editor of this book, I had Barbara R. Cochran, a teacher of French in a private school in a suburb of Cleveland, make the basic translation. Using my own knowledge of French (I had studied it for 7 years in high school and college and had lived in Paris for 2 years), I edited her draft, comparing every page with the original. This procedure had 2 advantages. It caught omissions and errors. And it eliminated many of the Gallicisms and infelicities that often creep into a translation because the translator is so close to the original. In addition, I tightened the text, corrected a few slips of the pen, clarified the occasional obscurities, and converted metric to English measure and the French cipher examples to English. I next submitted the edited portions of the section on radios to an expert on World War II radios, Will Jensby of Davenport, Nebraska, and of the section on guns to the editor of the *Shooter's Bible*, Robert F. Scott of Croton-on-Hudson, New York. They corrected technical terms that had been improperly translated, rendering them into their American equivalents. Their work has enormously improved the book, and I am very grateful to them. Then I sent the entire translation to Pierre Lorain, marking words and phrases that were not clear. He went over it, using his knowledge of English, making suggestions and corrections. During 2 of my trips to Paris, Lorain and I discussed the translation at length, clearing up the remaining obscurities, deciding how best to express difficult passages, and restoring the original English of quotations and of measurements from World War II manuals that he had in his possession. Finally, each of us read the proofs to eliminate any remaining mistakes. The result, author and editor believe, is a translation that, in the proper American technical terminology, accurately depicts its subject matter.

We consequently offer it to the English-speaking world in the hope that it will both instruct and please.

David Kahn

Great Neck, New York

# Foreword

by Brigadier F. W. Nicholls, C.B.E., M.B.E.

Communications and supplies are fundamental to the initiation and conduct of "special operations."

The English version of Pierre Lorain's *Armement Clandestin, France 1941–1944* jointly highlights this concept and expresses the gratitude of France for the unstinted provision with which both were made. It will give great satisfaction to its readers.

Those assembled by Major-General Sir Colin Gubbins, K.C.M.G., D.S.O., M.C., and his staff—the "backroom boys," designers and factory workers who produced the specialized equipment mentioned, and many others who produced the 1,001 items ranging from forged paper money to miniature submarines—will be proud to know that their contribution enabled overrun France to regain her self-respect, and that thus provided, the Resistance Forces so much influenced the course of the war as to bring it to an early conclusion.

Pierre Lorain can feel assured of a feeling in the hearts of all concerned of enduring gratitude.

# Foreword

by Gilbert Renault-Roulier (Rémy)

*. . . Politics then reigned in Rome. She had her two sisters, Deceit and Greed, as ministers. Under their command, Ignorance, Fanaticism, and Fury were seen to prevail in Europe. They brought wretchedness everywhere they went. Reason hid in the bottom of a well, along with her daughter, Truth. No one knew where this well was. If anyone had guessed, he would have gone down into it to cut the throats of mother and daughter.*

Voltaire, *Éloge historique de la Raison*, 1775.

Nearly 2 centuries later, these lines by Voltaire still have an awful resonance! If across the years these words continue to ring true, it is because adherence to the fundamental values on which our existence is based demands constant struggle.

During World War II, this struggle was the one in which our fighting men operated intelligence-gathering networks to the best of their abilities. In this work by Pierre Lorain, splendidly written and illustrated at the cost of extraordinary labor, I rediscovered many of the "tools" we were supplied with. *Armement Clandestin, France 1941–1944* is an important contribution to the dossier on the "combatants of the night," whose efforts ensured that Freedom was not defeated.

RÉMY

# Notes and Abbreviations

The technical terminology of the time is used here; thus "kilocycles (per second)" (kc/s) is used by the Allies and "kiloHertz" (kHz) by the Germans, although the 2 terms have the same meaning and the latter has become the official term.

To avoid confusing the reader with complicated and useless acronyms, the multiple divisions of Military Intelligence (MI) are grouped together under the general term "Intelligence Service" (I.S.), an appellation consecrated by the usage of the most highly regarded British experts.

Unless otherwise indicated, all weapons and equipment shown are British. Although a part of the American Office of Strategic Services (O.S.S.) was merged with Britain's Special Operations Executive (S.O.E.) during the war, our American allies were not in the least prepared for subversive operations. And although they provided valuable help to the Resistance during the last months of hostilities in France with a large number of parachute drops, the S.O.E. and its methods should be considered typically and entirely British.

The French Forces of the Interior, which formed the underground resistance in France, were almost exclusively supplied with British equipment while the regular French military units received regulation American equipment.

| | |
|---|---|
| AI-10 | cover name at the (British) Air Ministry for S.O.E. |
| AL | S.O.E. air liaison section |
| B.C.R.A. | Bureau Central de Renseignements et d'Action—a Gaullist agency formed, in January 1942, with a section dealing with clandestine operations |
| D | Section D of the Foreign Office—a unit to plan attacks on enemies by means other than the operations of military forces |
| DF | S.O.E. escape section |
| EH | Electra House—symbol for the office, originally located in Electra House, London, that evolved into P.W.E. |
| EU/P Section | S.O.E. section in charge of Polish networks in northern France |
| F Section | S.O.E. "independent" French section; cf. RF Section |
| GS(R) | General Staff (Research)—unit of War Office General Staff that concerned itself with guerrilla warfare |
| H Section | S.O.E. Italian section |
| I.S. | Intelligence Service—informal designation for British intelligence agencies |
| kc/s | kilocycles per second |
| kHz | kiloHertz |
| kilocycles | short for "kilocycles per second" |
| mc/s | megacycles per second |
| MI | Military Intelligence (British) |

| | |
|---|---|
| MI-6 | branch of Military Intelligence assigned to gain information about the enemy |
| MO-1(SP) | cover name at the War Office for S.O.E. |
| N Section | S.O.E. Netherlands Section |
| NID(Q) | cover name at the Admiralty for S.O.E. |
| P.W.E. | Political Warfare Executive—British propaganda organization |
| RF Section | S.O.E. Free French, or Gaullist, section; cf. F Section |
| R.S.H.A. | Reichssicherheitshauptamt, Reich Security Administration—a German joint government and Nazi party espionage, counterespionage, and police (detectives) organization |
| S.O.E. | Special Operations Executive—British organization for subversion and sabotage against the enemy, mainly in occupied territories |
| S.R. | Service de Renseignements—Intelligence Service, a secret intelligence branch to gain information of enemy activities; sometimes used to designate a network activity |
| T Section | S.O.E. Belgian Section |
| TR | Travaux Ruraux—Rural Works, a cover name for the French secret counterespionage organizations |
| TR-J | Travaux Ruraux Junior—another cover name, this one supposedly spoke in praise of rural works where youth could be a great help |
| X | S.O.E. German section |

# Preface

The French Resistance, both Phoenix and Proteus, escapes any rigid classification. Nevertheless, to facilitate its analysis, the following protagonists can be singled out:

- Civilians of all political tendencies and from all milieux grouped according to relationships, professions, or the services they could provide;
- Those in charge of escape networks for prisoners of war or Allied pilots shot down over French soil;
- The intelligence and counterespionage professionals of the French army, dispersed, but working "together" in clandestine operations;
- The B.C.R.A. agents of General Charles de Gaulle's Free French in London;
- French officers and soldiers who served in MI-6 of the Intelligence Service or in the F, RF, or DF sections of the S.O.E.;
- And finally, during the last part of the war, an increasing number of determined men and young insubordinates in the compulsory work details in Germany who made up the nucleus of an extremely important clandestine army.

All these elements had a common denominator: the British Secret Service, which embraced them, provided indispensable means of communication, arms, food, and first aid. In a word, they enabled them to act and survive.

The mysterious S.O.E. was the principal, if not the only, instrument of the Resistance.

The British government decided some years ago to have this organization's official history written up. Published in 1966 by Her Majesty's Stationery Office, *S. O. E. in France* is the authoritative work of Professor M. R. D. Foot, a historian who teaches modern history, then at the University of Manchester, now at the University of Sussex. Professor Foot, who had access to all the extant archives of this secret service, which had been created out of necessity, produced a unique and definitive work. In it he tells the story of the organization, its methods, the networks, and the men.

With this as a basis, it was our intention to describe the "tools" and to explain their operation.

We tried not to render this technical study too dry by adding illustrations and some photographs of original documents. This book is for buffs: we put our best into it without worrying about the amount of time we spent on it.

During our lengthy research, we had the privilege of

meeting some exceptional men and women. As the years went on, the quantity and quality of the documents which they were willing to entrust us with became such that this study became a veritable mission to preserve some information about the equipment and methods which played so considerable a part in the history of France.

We express our gratitude here to all these British and French friends, alumni of the S.O.E. and Special Service groups, who devoted themselves unselfishly to furnishing us with the rare original documents that were able to be preserved.

Without their help, such an undertaking could never have been carried out. We regret that all those involved cannot be mentioned here since they are so numerous. However, we should especially mention the following persons because their help was so important:

Great Britain:

- Brigadier F. W. Nicholls, C.B.E., M.B.E., director of communications, S.O.E.
- Colonel Maurice J. Buckmaster, O.B.E., commander, Section F, S.O.E.
- Squadron Leader Maurice Southgate, D.S.O., HECTOR network, F Section, S.O.E.
- Flight Lieutenant Charles B. Bovill, Chartered Engineer, officer in Aerial Operations Research Section, S.O.E., Station IX

France:

- Mrs. E. Balachowsky, MARCO POLO network, B.C.R.A., and PROSPER network, F Section, S.O.E.
- Colonel de Buttet, veteran of the KLÉBER net. Curator at the Army Museum, Paris
- Major Elly Rous (at the time, Captain François Serra), mission commander, first class, of submarine and parachute missions in occupied France. Head of BADEN-SAVOIE network TR-J
- Yves Saumont, in charge of radio stations for the MARCO POLO network, B.C.R.A.
- Major Jean Gompell (at the time, second lieutenant, French detachment to the Indies)

We should not forget our friends and neighbors from the Creuse region of the S.O.E. F Section FIREMAN circuit, nor their dynamic Irish radio operator, MICHELINE, Lieutenant Maureen Patricia O'Sullivan, M.B.E.

All of the weapons and some of the radio sets were studied in French and British museums. To the curators of these museums, who opened their abundant holdings to us so that we were able to study the minutest details of all armaments dropped by parachute, our thanks:

- The general director of the Imperial War Museum, London
- The grand chancellor of the Order of the Liberation, curator of the Liberation Museum, Paris
- Colonel Wemaere, curator at the Army Museum, Paris
- Military Engineer-in-Chief First Class Mercier, director of the National Manufactory of Saint-Étienne, department of the Loire
- My friends A. W. F. Taylerson, M.A., eminent expert on British armament, and R. Marquiset, member of the curatorial staff of the museum of the National Manufactory of Saint-Étienne

We should not forget the English amateur radio station, G3ZWH, in Kent, which placed itself at our disposal so we could conduct 2-way communications between France and England to test different S.O.E. radio sets. This enabled us to appreciate the extraordinary quality of these instruments, which started up instantly after a silence of 25 years. Finally, a big thank-you to all of our amateur radio friends who demonstrated their fellowship when they obtained data and, on many occasions, the rarest pieces of equipment for us.

Chapter I

# Introduction

Subversive warfare is as old as mankind. All conventional warfare carries within it the seed of secret and parallel activities aimed at forcing a decision by psychological or terrorist actions.

While conventional warfare is the setting in motion of an apocalypse, subversive warfare, much more subtle, never ends. During peacetime, its purpose is to destroy the morale of a potential adversary and to steal his secrets. During smouldering or declared wars, subversive warfare adds political upheaval and sabotage operations.

Without a doubt, the episode of the Trojan Horse is the oldest subversive action. Spies, plots, war stratagems, and political assassinations followed; as conventional warfare became more costly and more hazardous, the proportion of subversive actions consequently increased.

Surrounded by complete secrecy and known to only a small number of persons, subversive action becomes known to the public only by chance. It can only succeed as long as the masses against which it is directed are not warned of its existence.

Categorized by public opinion as "fictitious," its credibility was nil before World War II.

And still the master's hand governed the Nazi subversive war for more than a decade. After hysterical declarations, after insults to statesmen, words of appeasement followed. After promises, false oaths. After declarations of peace, brutal aggressions. After assurances of friendship, assassination. Rarely in history could such indecency have yielded the fruit of such victories and have found an audience as credulous as it was naïve.

Evil had dressed itself up in sumptuous finery to dazzle those it wanted to assail. To weakened, divided countries in social and economic crisis, Naziism offered strength, order, social harmony, the satisfaction of work, prosperity, large-scale urbanization, and the quality of German products.

The Nazi epidemic was spreading across Europe; by the time the diagnosis was made, it was already too late.

However, 2 organizations had uncovered the evil from the beginning: the French S.R. (Service de Renseignements, or intelligence service) and the British Intelligence Service.

With the 1938 anschluss of Austria, Great Britain foresaw the possibility, if a conflict were to break out, of engaging in political subversion operations in Germany in the hope of instigating riots. Two highly secret divisions were created in the Foreign Office and one in the War Office for this purpose. Their existence was known only to a very small number of initiates by their initials, respectively, "D," "EH," and "GS(R)."

For its part, the S.R. carried out a secret investigation which disclosed a substantial implantation of German agents on French soil. The identities of "potential traitors" were carefully documented. Finally, the analysis carried out on the German army led to sounding the alarm with the French high command: in vain!

The authorities were not able to take such pessimistic and alarmist reports seriously. In the higher echelons of command, the credibility of a so-called subversive war was indeed nonexistent: so the stage was set for it to succeed.

In less than a month, the rest of Europe, already contaminated, succumbed to the Nazi evil.

After the fact, it is interesting to demonstrate the mechanism of subversive operations, because even if evil's throat has been slit, it can reappear later in other forms and places. It is important to make an early diagnosis.

The subversive war the Nazis carried out in the years preceding the conflict can be broken down into 3 stages:

### Preparatory Stage

This stage consisted of a subtle infiltration into the country to be conquered, penetrating the press and political, artistic, and intellectual milieux.

This penetration was mainly carried out through bribery and blackmail. However, to persuade the idealists, Nazi Germany presented itself as the champion of pacifism, liberty, and justice. She defended the oppressed, preached the suppression of abuses, and attacked capitalism, trusts, and high finance.

She carefully prepared a support program for Germans residing in the countries she wished to conquer while depicting these individuals to the international public as minorities oppressed by plutocratic governments.

Finally, by politicizing her youth, Nazi Germany allied herself with an idealistic and fanatical force. By presenting herself as champion of movement, of modernism, and of socialism, she turned other states into reactionary and chauvinistic nations. By preaching peace, she turned other countries into warmongers.

### Action Stage

This stage was brought about through the infiltration of secret agents expert in the manipulation of the masses and charged with provocation that would lead to brutal repressions by the local authorities; the Nazi party thus gained martyrs assassinated by the Polish or the Czech police.

Clandestine armament of minorities banded together in self-defense militias quickly followed. Meanwhile, a part of the autochthonous population that had been conditioned by the Nazis defended the legitimacy of National Socialism and denounced the bankruptcies and abuses of their governments.

### Final Stage

This consisted of intimidation through diplomatic means, supported by a demonstration of force at the borders; "spontaneous" riots among the population and political assassinations took place at the same time in order that the adversary government capitulate under the pressures of a sector of public opinion and of armed groups, which welcomed their peace-loving invaders as liberators. (For France, it was a matter of "tourists in uniform.")

Afterward, a vote, as large as it was predetermined, sanctioned the *fait accompli*.

If intimidation did not work, the Nazi government felt compelled to place the coveted state under its "protection." But the international community was solemnly informed about the aggression the Reich had just suffered. False documents on real cadavers of fake Polish soldiers "proved" to the international press that it was actually Poland that had treacherously attacked Germany at the border radio station at Gleiwitz at the end of August 1939.

Finally, Nazi Germany's constant concern was to appear as a victim compelled to defend herself after having suffered the worst affronts without flinching. In this way, the masses, for the most part naïve and fair-minded, wished nothing more than to believe in Hitler as a knight-errant who preached the ideals of peace, liberty, and justice.

It was a small number indeed who were able to discern the merciless ideological character of the war and who foresaw the terrible years of massacre and enslavement which would fall upon all Europe.

On June 15, 1940, at the General Headquarters in Briare, south of Paris, General Joseph Georges, commander of France's northwest front, announced to army commanders that the government had started negotiating for an armistice.

As a consequence, the French Army Service de Renseignements was dissolved, but it resolved not to honor the Franco-German armistice. The S.R. was secretly reborn from its own ashes, and it decided to continue working in conjunction with the British Intelligence Service stations remaining in France. For its part, the French counterespionage group prepared for future operations under the anonymous and reassuring (for Germany) title, "Rural Works Service" (Service des Travaux Ruraux, abbreviated as TR).

However, England, all alone, was at bay. She needed to employ all available means to prevent the economic and moral capital of France from swinging toward the enemy camp. In London during July, the three mysterious departments of the Foreign Office and the War Office merged: the S.O.E. was born.

# Chapter II

# S.O.E.: "The Company"

"Set Europe ablaze!"

Such was the directive of Prime Minister Winston Churchill to his minister of economic warfare, Hugh Dalton, charged with running the newly created agency while at the same time disguising it.

This key phrase determined the organization's mission. It seemed gigantic: nothing less than sabotaging the factories and supplies of energy in territories annexed by the Germans, creating, well in advance, the prerequisites for an armed uprising of subjugated peoples, and finally, at the right moment, and with the help of the Imperial General Staff, bringing arms into the occupied countries in sufficient numbers to equip a secret army capable of fighting behind enemy lines until British troops could again land in Europe.

All signs led to the belief that Europe in general and France in particular still had enough healthy elements wanting to carry on the struggle.

The new organization was christened "Special Operations Executive" and designated by its initials, "S.O.E."

This name, however, was only known to a very small number of officers, and the very existence of this new service was one of the best-kept secrets of the war.

The Admiralty knew it only as a department called NID(Q), the War Office, as MO-1(SP) section; the Air Ministry maintained close contact with AI-10. The Foreign Office dealt with another unit. Laboratories might carry out research or build equipment for the "Joint Technical Board" or the "Inter-Service Research Bureau" without suspecting that all these names were covers for the unity of the formidable and omnipotent S.O.E.

Its secret headquarters, initially at 64 Baker Street in central London, soon branched out into nearby premises. Very often, S.O.E. agents saw nothing more of the organization they worked for, and which they called "the Company," than a hotel room, a small flat, or even an office located, for example, in London's Marylebone or South Kensington districts. Throughout the war, the Germans believed that a clandestine organization working on France was headquartered in a London apartment.

Baker Street began operating with some 10 telephone lines; by 1944, it had more than 200. These lines ran to 4 large secret radio stations hidden on premises some distance from London, at Grendon, Poundon, and Bicester, all in England, and at Dunbar, Scotland. These were designated Stations 53-A, 53-B, 53-C, and 53-D, respectively. The lines also went to secret laboratories, where mysterious equipment was developed, to extremely discreet schools, where one learned many things which could not be discussed, and to a camouflaged airfield, to which the Royal Air Force had generously detached 2 squadrons, the 138th and the 161st, made up of first-rate navigators and pilots entrusted with the secret transport of men and materials.

Finally, a little flotilla of 5 innocent-looking fishing boats, equipped with trawl nets and tuna nets, was stationed on the Helford River west of Falmouth. A corner of the harbor of Gibraltar sheltered a 1,500-ton cargo ship and two 20-ton motorized feluccas.

There were 4 S.O.E. divisions working in France.

The "independent" F Section was the largest in terms of numbers.

The RF Section was charged with working more particularly with the secret agents of Charles de Gaulle's Free France. This section trained the agents, supplied them with the necessary equipment, inserted them into France, and brought them back out.

The EU/P Section was in charge of Polish networks in northern France.

Finally, the DF Section was charged with setting up escape channels for agents returning to England.

For obvious security reasons, no section was acquainted with the one next door, and each worked according to its own methods, certain that these were the best.*

A section designated AL served as the communication and liaison office with the Air Ministry.

A section designated "Stocking-Packing" prepared cases for delivery.

Finally, an organization separate from the S.O.E., called P.W.E. (Political Warfare Executive), was in charge of propaganda aimed at Germany and occupied territories.

*In the rest of Europe, N Section was in charge of Holland, T Section of Belgium, X Section of Germany, and H Section of the Iberian Peninsula. Other sections worked in Norway, Czechoslovakia, the Balkans, the Middle East, and the Far East.

At the end of 1942, part of the American O.S.S. was integrated into the S.O.E.

Ultimately, all these sections were subordinated to the British Chiefs of Staff Committee and the War Cabinet.

It is important to remember that the mission of the S.O.E. was subversion, sabotage, training, and clandestine arming of guerrillas and that it excluded all intelligence gathering. The latter was carried out in France by the Intelligence Service (MI-6) in conjunction with the B.C.R.A. of the Free French on the one hand, and the special secret service of the French army on the other.

This fundamental difference in assignments did not make relations between the S.O.E. and the I.S. easy.

The agents of the I.S. needed an environment as close as possible to that of peacetime to accomplish their mission. Their activities could only be carried out if the German police were off guard. Thus, S.O.E. activities which aimed at creating a climate of civil war could only appear adverse to them, for they risked drawing attention to a service whose main characteristic was supposedly absolute invisibility.

Finally, traditional organizations would only grudgingly give in to S.O.E. methods, which were often contrary to the laws of war.

Because of all this, the S.O.E., at least during the first years, met with blind opposition among the more orthodox services.

In any case, if the missions of the S.O.E. and the I.S. differed, the technical means, the special weapons, and the communications systems were identical.

We shall attempt to classify and describe them in this book,

# Chapter III

# Intangible Links: Communications and Secret Radio

# CONVENTIONAL RADIO TRANSMISSIONS

### Definitions

A radio transmitter is an instrument that produces high-frequency electromagnetic radiation. By means of a radiator—an antenna or aerial—it radiates these waves, called "Hertzian waves," through space at the speed of light.

A radio receiver, located at any point in space, picks up these waves with an aerial, selects, from among them, the frequency the operator has chosen, renders it audible in detecting it, and finally amplifies it to a suitable level.

A transmitter is characterized by:

1. The frequency or range of frequencies on which it can transmit. Frequencies are expressed in kilocycles or megacycles.
2. Its power; in other words, the energy that it can emit through an aerial. Power is expressed in watts.
3. The type of signals transmitted: telegraphic, telephonic, televisual, pulsating, etc.

A receiver is characterized by:

1. The frequency or range of frequencies it can receive.
2. Its sensitivity; in other words, its ability to pick up weak signals.
3. Its selectivity or bandwidth; in other words, its ability to separate 2 different transmissions of neighboring frequencies.

Power is not listed as one of the main characteristics of a communications receiver; there is always enough to operate headphones. Finally, transmitters and receivers are both characterized by frequency stability, that is, the ability not to drift from the chosen frequency during transmission or reception.

Since transmitters and receivers are electronic instruments, they need an adequate power supply to function: batteries, accumulators, converters, generators, or feed circuits.

A transmission station will have, at the least, a transmitter, with a power supply and an aerial, and a receiver, with its power supply and aerial. Accessories might include a cipher machine, a teletypewriter, or high-speed automatic telegraph equipment. At the lower limit, a low-power station will have a transmitter, a receiver, a power supply, and an aerial shared by both. This ensemble, which can be very compact, is a transceiver.

### Frequency Range

It is extremely easy to produce a radio wave. All one needs to do is send an alternating current of sufficiently high frequency through a circuit. As an example, let's take an alternating current with a frequency of 50 cycles per second (that is to say, it oscillates 50 times per second) produced by the house current. Raise the frequency to 50,000 cycles per second; in other words, oscillate the frequency 1,000 times faster than before. The circuit will then transmit a radio wave with a frequency of 50,000 cycles (or 50 kilocycles) in all directions at the speed of light.

If this high-frequency current is interrupted in the rhythm of Morse code, radiotelegraphic signals will be transmitted.

A continuous wave, carrying a voice (radio telephony) or an image (television), can also be transmitted.

Hertzian waves are arbitrarily classified as follows:

Long waves (frequencies of 50–500 kilocycles per second [kc/s])
Medium waves (500–1,500 kc/s or 1.5 megacycles/second [mc/s])
Short waves (1.5–30 mc/s)
Very short waves (30–200 mc/s)
Ultrashort waves (with frequencies greater than 200 mc/s)

Beyond 1,000,000 megacycles are found, in succession, infrared rays, visible light rays, ultraviolet rays, and X rays.

International agreement strictly regulates the assignment of frequencies. Long and medium waves are reserved almost exclusively for radio broadcasts. Short waves and very short waves are distributed among land, sea, and air services, with some bands set aside for radio broadcasts, television, and amateur radio operation. Ultrashort waves are used in air navigation, by radar, for microwave beams, and by amateur radio operations.

## Power Range

In theory, a Hertzian, or radio, wave is propagated equally in all directions. In practice, however, the geological and geographic configuration of the transmission site, and the presence of natural and artificial obstacles (hills or metal constructions) between the transmitter and receiver, inducing more or less absorption of transmitted radiations, could reduce or even block propagation in a certain direction, even though these same obstacles might reinforce the signals in the opposite direction.

Generally speaking, as the signals move farther away from the transmitter, their intensity gradually diminishes until they die out completely. An increase in transmitter power and receiver sensitivity, joined with the utilization of aerials with high gain, extends the boundary of the signal's "zone of extinction."

Because of the preponderant influence of the surrounding area, it is not possible to establish a definitive relationship between power and transmission range. Nevertheless, to form some idea, one could say that for radiotelephone transmitters operating from 2 to 30 megacycles per second:

With 0.5 to 1 watt of power, the average utilizable transmission range is in the area of 1 to 2 miles.

With 10 watts of power, this range could extend to approximately 10 miles.

With 100 watts of power, a range of 25 miles may be counted on.

Finally, for a large, 1-kilowatt transmitter, a range of 35 miles may be expected.

Using these same transmitters for telegraphy, the range doubles: approximately 25 miles for 10 watts, 50 miles for 100 watts, and 100 miles for 1 kilowatt.

This propagation, said to be by ground wave since the wave produced is diffused very close to the ground, allows reliable and unbroken communication between the transmitter and the receiver, at any time and under any circumstances. In radiotelegraphy links, these signals extend almost to the boundary of extinction. This is the type of propagation that best suits classical military applications.

## The Principles of Radiolocation (Radio Goniometry, or Direction-Finding)

Owing to the very fact that a transmitter radiates, in all directions, an electromagnetic field whose intensity grows weaker as it travels farther away, radiolocation becomes immediately possible once 2 or 3 receivers are set up in the field covered by the ground wave.

Let's assume that there are 3 receivers, *A*, *B*, and *C*, all located in the field of transmitter *T*. Each receiver is equipped with rotatable, high-gain loop aerials that present a large absorption surface when their greatest dimension faces the source of the radiations.

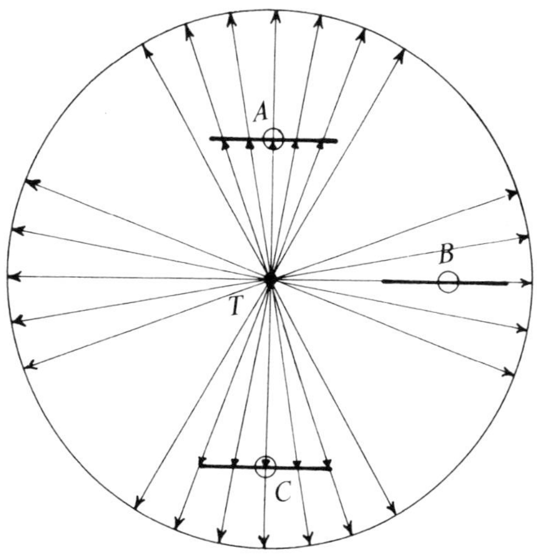

The diagram shows that the antennas of receivers *A* and *C* will receive most of the radiation while the antenna of receiver *B* will pick up almost none. Signals received by stations *A* and *C* will be at maximum power while receiver *B* will remain silent. If the rotatable antennas are connected to a dial forming a compass card, the location of the transmitter will be instantly approximated by both the *A* and *C* stations. At this point, the *B* station, by turning its antenna to where it hears the signals at their loudest, will confirm the transmitter's location within a radius of several hundred yards. A vehicle will advance into the area thus circumscribed. Signals given off by a simple device measuring the intensity of the field emitted by the transmitter will automatically lead to the place where the intensity of the field is greatest. This would be the exact spot from where the signals are transmitted.

## Long-Distance Propagation

Ground waves whose range is relatively limited and that rely heavily on the power diffused by the aerial are not the only kind of wave that a radio transmitter emits. The transmitter simultaneously puts out an aerial wave which is projected toward the upper atmosphere.

At an altitude of 60 to 200 miles, this wave will meet an ionized zone called the "ionosphere." It will reflect off the ionosphere, like light off a mirror, if its frequency is approximately 2 to 30 megacycles. The wave will return to the earth, only to bounce back to the ionosphere. Thence, reflection after reflection, it will travel around the globe, leaving silent zones of varying size between each rebound.

We have seen that because ground waves travel extremely close to the ground, they are continuously and rapidly absorbed. We have also seen that this can only be partially compensated for by increasing the transmitter's power. On the other hand, aerial waves are (partly) absorbed only at points of contact with the ionosphere and the earth. Therefore, a signal received from several thousand, or even tens of thousands of miles away, will often be much stronger than the same signal picked up by a ground wave a mere 20 miles from the transmitter.

Power thus is no longer the determining factor of range. A 10-watt transmitter whose ground wave, under the best of conditions, dies out at 10 miles in telephonic and 25 miles in telegraphic transmissions, can carry out excellent communications between Paris and South America with ionospheric propagation.

(This kind of propagation is impossible for frequencies greater than 30 megacycles. Beyond this frequency, the aerial wave does not reflect off the ionosphere, and only the ground wave can be employed. Therefore, ultra-short waves and very short waves can only be used for line-of-sight communications or communications that reach just barely beyond the horizon.)

Ionospheric propagation was discovered by amateur radio operators who, in 1923, completed the first 2-way communication between the United States and France on a frequency of 2,700 kilocycles. They opened the way for intercontinental telecommunications, which, after 1930, reached a high level of efficiency.

Ionospheric propagation permits long-range goniometric direction-finding, beyond borders, under the same conditions we looked at for short-range direction-finding. The only differences would be that since long-range radiogoniometer stations are unable to enter the ground-wave field, localization will be less precise, and the last stage, or the advance with the help of the field-strength meter, will obviously not be able to take place. The silent zones between each bounce vary in size from about 100 to 2,500 miles, depending on the frequency employed. They also fluctuate with the time of day, the time of year, and the solar cycle.

Each reception area can range between about 50 and 300 miles.

Regularly updated charts called "ionospheric propagation forecast charts" indicate for the coming months, in terms of transmitter location and the communication to be carried out, the frequencies to be employed and the times at which communication can be conducted with a probable success rate greater than 75 percent.

Therefore, through a wise choice of hours and fre-

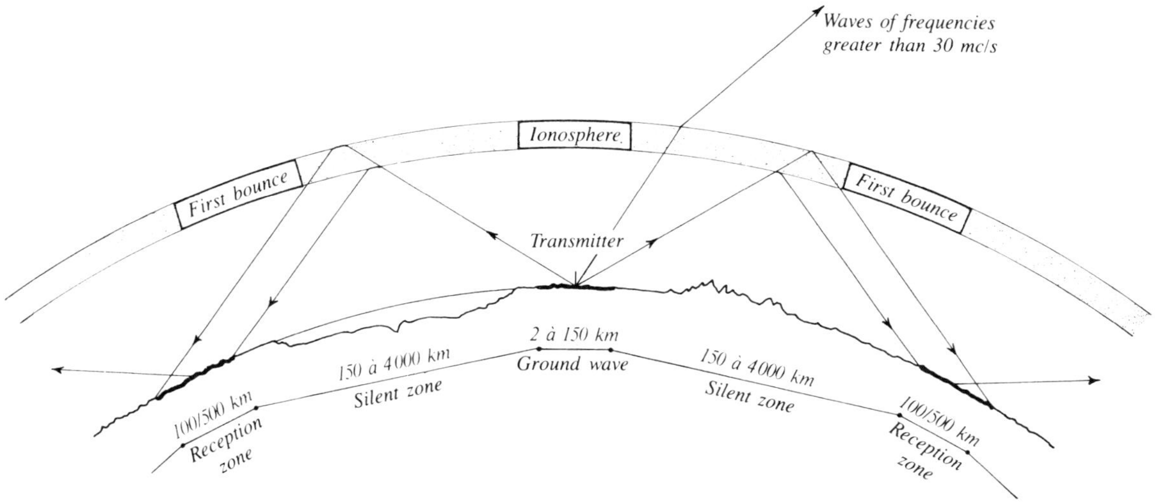


quencies, a warship can travel around the world while remaining in almost constant communication with the mother country.

The use of powerful transmitters and high-gain antennas will somewhat increase the number of hours of communication on a given trip. In theory, however, power on the order of a single watt, radiated through a properly selected aerial, will permit telegraphic communication with the antipodes if the frequency is wisely chosen.

## CLANDESTINE RADIO TRANSMISSIONS

### The Imperatives of Frequencies

We will address ourselves to the problem faced by the S.O.E. upon its creation: how to maintain 2-way radio communication between occupied Europe (mainly France) and its London home station. This represented a minimum range of 100 miles for a clandestine station located in the Pas de Calais department, on the Channel opposite Dover, and a maximum range of 1,000 miles for a station operating in the most distant occupied countries. Moreover, communication had to be maintained day and night.

Frequencies permitting coverage of these distances with a single ionospheric reflection ranged from about 3 to about 8 megacycles. For France in particular, tests showed that the optimal frequency between 8 A.M. and 6 P.M. was approximately 6 megacycles; between 7 P.M. and 7 A.M., approximately 3 megacycles. If the clandestine station was located on France's Mediterranean coast, a frequency around 8 megacycles facilitated daytime communications.

### The Imperatives of Power

Since the range would be covered with a single bounce of the aerial waves, power was not a predetermining factor. It was rather the imperatives of weight and space that necessitated choosing very low power transmitters of 1 to 10 watts. This also contributed somewhat to protection against enemy radiolocation by reducing the extent of the ground wave.

By the 1940s, a 1-watt transmitter could be very compact: the size of a small cigar box. With a properly erected outdoor aerial, it could radiate a signal that could be heard loud and clear in London.

### The Imperatives of the Aerial

Without a doubt, a properly mounted outdoor aerial was the first thing which had to be completely banished from a clandestine transmission station because there was a good chance it would attract the eye of an expert or intrigue a merely curious passerby.

If hurriedly set up in the housemaid's chambers or an apartment, a bent and tangled clandestine aerial would beam most of the transmitter's energy into the surrounding walls. In addition, the tuning of a set that was hastily set up was rarely precise; accordingly, energy loss could reach very high proportions.

If a 1-watt transmitter was used under these conditions, the energy effectively diffused in space was most likely insufficient for communication to take place.

On the other hand, a 10-watt transmitter operating under the worst conditions might still have been able to radiate 2 or 3 watts. This permitted effective communication without emitting too strong a ground wave.

### The Imperative of the Type of Signals to Be Used

Conventional military transmissions used radiotelephones in the lower echelons for communications between vehicles, air force pursuit, reconnaissance units, and anywhere increased voice range was needed for direct orders requiring immediate attention. As a rule, these transmissions were in clear—that is, not enciphered—though they were sometimes disguised in a very simple manner.

In the higher echelons, communications were conducted through enciphered telegrams, transmitted by automatic telegraphy, teletypewriter, or facsimile transmission systems. Manual telegraphy was employed for difficult communications.

Clandestine transmissions were a lot less accommodating. To start with, all heavy, cumbersome, noisy, high-electricity-consuming, and nonessential equipment was set aside—in other words, all automatic transmission equipment.

It must be remembered that an agent working in an enemy country was encumbered even with the strict minimum that a transceiver comprised.

This left only the radiotelephone and the radiotelegraph from which to choose. This choice was also easy to make. The radiotelephone required a higher level of power for a shorter range, was particularly indiscreet, facilitated radio position-finding, was easily garbled, and required repetitions and the use of the phonetic alphabet to spell out each letter in secret messages that were to be transmitted. In other words, the radiotelephone clearly required more time than radiotelegraphy to transmit a given message, and the result was infinitely less certain.

The following diagrams summarize the differences between voice transmission and Morse signals.

Figure 1: A radiotelephone signal requires a minimum bandwidth of 6 kilocycles per second to be comprehensible (in amplitude modulation, the mode of transmission most used at the time) whereas a Morse signal needs only a width of 0.2 kilocycles (or 200 cycles). The diagram shows that a transmission on radiotelephone *A*, which is centered on 7,000 kilocycles, will actually congest 6,997–7,003 kilocycles. A transmission on radiotelegraph *B*, centered over 7,004 kilocycles, shows a negligible congestion.

Figure 2: The three radiotelegraph stations *B*, *C*, and *D*, transmitting on 6,998, 7,001 and 7,003 kilocycles respectively, can transmit simultaneously without interfering with one another. The station transmitting by radiotelephone *A* on 7,000 kilocycles will be overwhelmed by Morse signals to the extent that it will be totally incomprehensible.

Even during World War II, the crowding of the shortwave spectrum threatened to prevent any effective radiotelephone transmission by low- and medium-powered equipment, except over very short distances. Accordingly, around 1942, most stations transmitting by radio-

Fig. 1

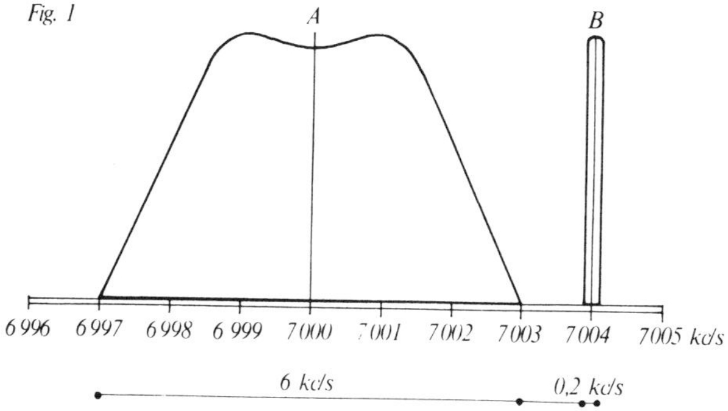


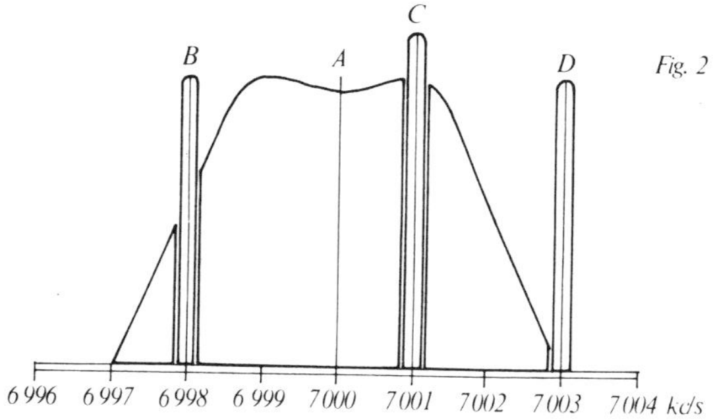


Fig. 2

telephone, especially for reconnaissance and tank communications, shifted to very short waves, in other words, to frequencies greater than 30 megacycles. Thus, they had very many short-range channels of communication that were free of interference.

Signals specialists no longer had a choice. The manual radiotelegraph imposed itself, and the obligation arose of training excellent radio operators to transmit Morse at a minimum speed of 20 to 25 words a minute, or 100 to 125 letters or digits per minute.

### The Imperative of Stability

The clandestine transmitter was generally heard at the home station with a weak signal, and since it could only send during an extremely short period because it had to evade enemy direction-finding, its frequency had to be fixed with absolute precision so that the home station would not lose time trying to locate it. The easiest and most efficient way to transmit on a fixed frequency and ensure perfect stability was by using a quartz crystal.

Briefly, a slice of quartz crystal the size of a postage stamp, when stimulated by a high-frequency current, maintains highly stable vibrations whose exact frequency is determined by the thickness of the thin crystal plate. High-precision machinery could cut quartz for any given frequency, almost within a kilocycle.

A clandestine transmitter had a minimum of 3 crystals for day, night, and emergency frequencies.

### The Clandestine Receiver

It did not differ from a regular communications receiver except for its compact size. Meant for the reception of Morse signals, it had to have as great a selectivity and stability as technology would allow. It was highly sensitive, but its power was limited to the maximum for headphone listening.

### The Power Supply

Common to both the transmitter and receiver, the power supply as a rule permitted using house current of any voltage, as well as a simple car battery. It was of especially strong construction, and at the same time, as compact and lightweight as possible.

With these ideas on transmissions, we can now start our discovery of these extraordinary and extremely mysterious small machines, today infinitely rare and precious.

Indispensable tools, the sole and slender links between the distant oasis and the arid desert of occupied countries, the secret S.O.E. transceivers were probably the most meaningful and most emotionally charged objects of the war. We should think of them with profound respect, for, in keeping a network alive, they often led their operators down the ways of death.

This is why at the beginning we must take a quick look at the friendly home stations carefully hidden in England, weaving in space invisible networks of high-pitched signals, transmitting in rapid insectlike chirps inexpressibly secret voices bearing death sentences for Germans and traitors, but also comfort and announcements of important events. And then, we shall describe the formidable spider's web that the German services extended to trap these voices, inaudible to the uninitiated and heavy with menace, to try to comprehend them in order to capture them, extract their secrets, and still them forever.

## THE HOME STATION

If a network's most valuable possession was a radio set (and there is no doubt that these instruments were worth more than their weight in gold), it was just as important to have a qualified operator. We have seen that, since radiotelephone communications could not serve, it was necessary to specially train radiotelegraph operators of exceptional quality. London never turned out enough of these operators, and as a result, there was an overload of work for stations that were already set up, to the detriment of their safety and the advantage of the German radiolocation forces.

For the clandestine operator, who lived under considerable nervous tension (when he left on his mission, it was not realistic to give him more than 3 months to survive), the home station represented at once his homeland; a leader whom he knew, admired, and respected; and a father, both tyrannical and benevolent, who encouraged him, sustained him in the face of adversity, advised him, sometimes even congratulated him.

So it was only natural for the radio operator to assimilate the home station to the tiny little cell he knew of the S.O.E. For obvious security reasons, the radio operator could not know what in reality this home station was nor suspect the number of sets and personnel assigned to his humble self, whose only job was to maintain contact.

The war years were characterized by enormous growth in all areas. An organization like the S.O.E., which was born, lived, and died within 5 years, underwent continual changes which during peacetime would have required more than 20 years to be realized. Each mission was an opportunity to try out new methods and new equipment, or to modify those used previously. Communications procedures only improved as they became more simplified, offering much more safety to an operator lost in a hostile world.

In May 1941, 2 clandestine stations were operating in France for Section F. In May 1942, there were still no more than 7. However, by December 1943, the number had almost quadrupled, and by August 1944, 53 clandestine networks transmitted a high volume of ciphered telegrams exclusively for Section F. This kept several hundred radio operators, cipher clerks, and teletypewriter operators going in the 4 big receiving stations located in different areas of Britain.

Between the inception of the S.O.E. and its dissolution, 2 great periods can be singled out. June 1, 1942, was the turning point.

Before this date, even though the S.O.E. was regarded by certain people—and not unimportant ones—as headed for a great destiny, it was judged by many others who did not know it as an organization of dubious utility, composed of inexperienced amateurs who carried out tasks which it was better not to know about. Accordingly, the S.O.E. often had to face the veiled hostility of several well established services that were very jealous of their prerogatives.

The first hurdle was to gain some means of radio communication: strictly speaking, some specialized sets could have been given up by the secret service; however, frequencies were scarce, and the organization in charge of allocating them was not very willing to do so if not told what they would be used for. Moreover, speaking of the creation of a secret radio network meant creating a receiving center. But such a center could not be thrown together within a few weeks. It would also be necessary to establish a cryptographic service for deciding which systems would be used and for distributing the systems and their keys. This would require highly qualified experts, who were not to be found within the newly created S.O.E.

For all these reasons, an agreement was made with the Intelligence Service, which had long had completely equipped radio networks and a cipher service employing the best experts. In this way, the secret services supplied the sets, the quartz crystals, the codes, and the operating schedules that the S.O.E. needed, and took over reception in their central.

From the technical point of view, this was a good solution, since it supplied the S.O.E. with all it needed. However, this alliance could not last, first, because the S.O.E. wanted its freedom, and then because the I.S., which had to deal with operating an increasingly dense network for itself, could no longer assume responsibility for the radio receptions of an ambitious and competitive secret service whose needs were growing in geometric progression.

On June 1, 1942, after more than a year of breaking in, the S.O.E., which had proven its usefulness and effectiveness, was finally authorized to assume full responsibility for its radio links. It could now build its own sets according to its own specifications, devise its own cryptosystems, and have a home station completely its own. This first home station would quickly add several branches as traffic grew.

Let us try to describe a typical central as it might have been at the beginning of 1944.

### Installation

The home station had to be set up in the country, far from London and its suburbs, for the clearest reception possible. A quiet Buckinghamshire estate surrounded by high walls will serve the purpose. Requisitioned on the cover ground that it is to serve for evacuation from air raids by a branch—an imaginary one—of some agency, it is discreetly guarded by the special services. A large number of young women go in and out during office hours, but others may enter only by invitation, and this is never tendered.

### Connections with Baker Street

This is instantaneous, by scrambler telephones and teletypewriters on special lines. These are protected in such a way that any interception that might be made on the line itself would be totally incomprehensible.

### Communication with the Clandestine Stations

On the grounds, far away from prying eyes, aerials standing in arcs face from northern Norway to southern Spain. Inside the buildings, there is the quiet atmosphere of a library and about 50 receivers constantly on alert. The operators are wearing headphones, pens and note pads ready.

These highly sensitive and selective receivers can, because of their bandwidth, which can be narrowed to 50 cycles, isolate and pull out the most tenuous Morse signal from a static from which it seems inextricable.

A wire recording device is attached to each receiver. This supplies an objective recording of the communication and ensures that not a single letter is lost. This recorder, usually at rest, is started by a radio operator as soon as a signal is heard.

But it serves another purpose, too. If Baker Street doubts the true identity of the correspondent—if, in other words, Baker Street fears that the outstation has fallen into enemy hands and that the enemy is using it to mislead the home station—the recording will verify the clandestine operator's identity by a simple comparison with a sample tape he recorded before leaving on his mission. For no operator, no matter how experienced, can produce Morse signals with the regularity of a machine. He makes very slight mistakes, such as a dot a little too long or a dash a little too short, always in the same letter; or he will tend to run the dots and dashes together in another letter. These shortcomings, which do not stand in the way of comprehension, constitute the personality of the radio operator's Morse writing, and accordingly, are veritable fingerprints characteristic of and peculiar to each operator.

### Calling Schedules

In 1944, Section F attained such a dense and high volume of traffic that each clandestine radio operator had to be supplied with a very strict operating schedule. Each microfilmed schedule was given to the operator before his departure or was dropped by parachute during his mission. It listed the days of the week and the hours when the home station would listen in on the frequencies allocated to him.

These frequencies were generally set up with 3 different quartz crystals. Sessions normally took place 3 times a week. The home station listened to each frequency for 5 minutes on the day and at the time established. Accordingly, the clandestine station no longer had to send out a call; knowing the home station would be listening, the outstation sent its message immediately.

It was specified that the length of a transmission on a given frequency could not exceed 5 minutes under any circumstances. This corresponded to a message of 120 groups (600 letters). As a rule, it was advisable to set the limit at 100 groups. Finally, the length of total transmissions in a single day was never to exceed 20 minutes.

Moreover, to cut to the minimum the amount of time an operator spent at his transmitter, it was decided that daytime hours would be exclusively reserved for transmissions from occupied countries while all transmission from the home station would take place at night.

Therefore, the operator only remained at his transmitter for the time needed to send messages. He separated the transmitter unit from the receiver unit and set up the latter in his bedroom. Or, better, he would install a new, specially parachuted receiver. There, with all doors bolted and with no fear of being discovered, the operator could easily take down messages from the home station and decipher them.

This arrangement enabled the home station to prepare on punched tape all telegrams arriving by teletypewriter from Baker Street during the day and to transmit them *en bloc* by automatic radiotelegraph between 9 P.M. and midnight. This ensured that perfectly rhythmical signals would be transmitted at an adequate and steady speed, so that the correspondents would not be tired needlessly.

Furthermore, this completely depersonalized all transmissions from the secret home stations, thus depriving the enemy of an extra chance to analyze the messages.

### Emergency Procedure

If the network had an urgent message to relay that could not wait until the next session, its radio operator could use one of his 2 quartz crystals for emergency communications. One established a daytime frequency; the other, a nighttime frequency.

The home station constantly monitored these frequencies. The agent was sure to receive a response to his emergency call exactly 1 hour and 10 minutes after he sent it.

This procedure was very risky since the clandestine radio operator had to contact the home station to be sure it had received the message correctly. In a sample case, he confirms this through a rapid exchange of service calls:

> BLK QTC 1 QRK? K [From BLK (that day's callsign for the clandestine station). QTC 1 = I have 1 message to send to you. QRK? = What is my signal intelligibility? K = Over.]

> BLK QRK QRV K [To BLK. QRK = Your signal is intelligible. QRV = I am ready. K = Over.]

The operator then taps out his message. The home station acknowledges receiving it and asks the operator to stand by:

> BLK QSL AS [To BLK. QSL = I acknowledge receipt. AS = Stand by.]

Behind the locked door, the radioman has to wait 70

minutes for the response, without budging. His eyes are fixed on the road, anxiously looking out for the sinister black Citroën. His Colt pistol, a cartridge in the breech, lies near his hand. His ears are alert to the slightest noises. Seventy minutes becomes an eternity.

And nevertheless, they work very fast in England.

When the central's operator transmits QSL AS, the clock stands at 10:30 P.M. The sheet is torn off the pad and sent to the cipher room. A cipher clerk enters the cipher groups under the key, and with the help of the alphabet table, restores them to plaintext. A 30-group telegram requires 20 minutes for deciphering: it is 10:50 P.M.

The deciphered message is typed on the keyboard of a teletypewriter; at the other end of the line, 30 miles away, an identical machine begins to run. The message arrives at Baker Street at 10:55 P.M.

Only 20 minutes are left to grasp its meaning, decide on a reply, and draw it up. At 11:15, an operator at the home station turns on his teletypewriter. The machine raps until 11:20. The operator tears off the typed page and hands it to the young cipher clerk. She breaks the message down into groups of 5 letters, writes these groups under a new key and converts it into cipher using her alphabet table. At 11:40, she has the cipher groups given to the radio service. The operator immediately taps out: BLK QTC 1 QRV? [To BLK. QTC 1 = I have 1 telegram for you. QRV? = Are you ready?]. The response is immediate: QRV K [Ready; over.].

The groups are transmitted manually so as not to lose time and as quickly as possible. It is 11:42 when the home station hears the acknowledgment from the clandestine station: QSL AR [QSL = I acknowledge receipt; AR = Over and out]. Everything has gone well.

The radio operator hastily folds up his aerial and his ground wire, removes the quartz crystal, disconnects the headphones, locks his suitcase, and secures his pistol. He slides everything back into its hiding place and rolls the cipher message up and inserts it into the handlebars of his bicycle. It is 11:47 when he sets out into the night, alone.

## GERMAN DIRECTION-FINDING

There is, at the same time, a central in the Paris region as secret, exclusive, and efficient as the London home station. Although it has the same cloistered library atmosphere, the equipment is different. Attached to each receiver are a televisionlike cathode-ray tube and a telephone. Three hundred receivers hum softly while 300 screens emit a cold greenish glow as they constantly monitor 30,000 frequencies between 10 kilocycles and 30 megacycles.

These ultramodern receivers are called "panoramic" receivers. This means that each cathode screen gives an instantaneous picture of all radio traffic on a band 100 kilocycles wide. Each station shows up as a luminous spot. The screen is graduated in kilocycles, and each operator has a list of frequencies used by stations that are under German control or are abroad. Therefore, any spot suddenly appearing outside of these known frequencies unquestionably represents a suspicious station, which very likely is clandestine.

It is 10:25 P.M. A spot has just lighted up on one of the screens. Someone takes down the telephone, announces slowly and clearly, in a low voice: "*Achtung, drei vier sechs funf KiloHertz*" ("Check three four six five kilo-Hertz") and hangs up. The other hand sets the Telefunken receiver Köln *A* to a frequency of 3,465 kilocycles while a wire recorder begins to hum.

We are no longer "at home," but at the very center of the spider's web of the formidable Reichssicherheitshauptamt, a secret organization often designated by its initials R.S.H.A., whose Department IV, the Gestapo, is in charge of reprisals against subversive plots.

Telephones have rung simultaneously at the same 10:25 P.M. in the large goniometric stations at Brest, on the tip of the Brittany peninsula, and at Augsburg and Nuremberg, both in southern Germany. After receiving the message "*Achtung, drei vier sechs funf KiloHertz,*" an expert at each of the stations sets the frequency on a Telefunken. A round cathode-ray screen, with a compass rose marked on its periphery, is attached to the set. A lighted beam instantly springs out from the center of the screen, indicating an extremely precise azimuth, or direction.

It is 10:27 P.M. Brest, Augsburg, and Nuremberg relay their 3 azimuths to the operations room, where they materialize as threads on a huge wall map of Europe, extending from the 3 goniometric bases and crossing to form a triangle with sides about 10 miles long.

The unknown station is located in the operations zone of the Clermont-Ferrand mobile regional base in central France. It is 10:34.

The operations room radios the Clermont base and gives the suspected frequency and the position-finding coordinates. Clermont acknowledges the message and calls the garage. Two front-wheel-drive Citroën 11s, each equipped with 4 civilians carrying machine pistols, and two 4-seater Mercedes-Benz convertibles with fake French license plates, dash off.

The receivers inside the vehicles are quickly set on 3,465 kiloHertz. It is 10:39. The clandestine radio operator ended his transmission 9 minutes ago. The frequency is quiet. Only a teasing background noise reaches the receiver headphones.

Nevertheless, the vehicles move toward the intersecting points of the position-finding triangle, lie in wait at road crossings, and stay on the lookout. They constantly communicate by radio among themselves and with their command post, which has just signaled that the unknown is waiting for a reply from his home station. If the clandestine operator had incautiously communicated again with his base, he would have been done for.

The technical teams in the two Mercedes wait anxiously. At 11:40, the response from London arrives. The rotating aerials are ready for a final survey. The Citroëns rev up their motors for the hunt.

At 11:42, the clandestine outstation replies: "QSL AR." The transmission lasts less than 3 seconds—not enough time to determine the station's position. The cars return empty-handed.

From this point on, however, Clermont has this frequency surveilled by patrol cars prowling the edges of the suspect triangle. With any new transmission from the clandestine outstation, the cars would situate their prey within a new, smaller triangle with sides half a mile long.

Meanwhile, the magnetic wire which recorded the messages is sent to the cryptanalytic service. With the help of electromechanical equipment, the service determines within a few minutes if the texts are in an unbreakable system or not. If the messages cannot be solved, the order is sent to Clermont to liquidate the outstation immediately. This outstation is extremely dangerous and nothing can be learned from analyzing its traffic. On the other hand, if the services uncover some weaknesses, the code will undoubtedly be broken within the coming days or weeks. Clermont is ordered to watch the transmission site discreetly and allow it to operate for the time being so that, when the time comes, the entire network can be hauled in.

If the R.S.H.A. is relentless and the radio operator persists in transmitting from the same place, he soon becomes the victim of a man who is looking at the watchlike dial on his wrist of his field-strength meter.

When the suspect area is reduced to a triangle of 200 yards on a side by the Mercedes, to which has been added, for closing-in operations, a small delivery truck covered with a tarpaulin, these vehicles are replaced by some harmless-looking pedestrians wearing bulky raincoats to hide their reassuring *bon vivant* stoutness. These pedestrians appear to be late, because they walk quickly while looking at their "wristwatches." Their stoutness conceals an extremely sensitive field-strength meter. The dial of the wristwatch is a measuring scale whose hand movements tell a Gestapo agent whether he is getting warm or cold. Two or 3 men, each with one hand in a pocket, follow each pedestrian. One pedestrian suddenly disappears into a building. The group marches after him. A car stops next to the sidewalk . . . the radio operator is done for.

If he had time to swallow his poison tablet, the removal of the body is very discreet. If the tenants notice anything, the men say that their neighbor has just been arrested for dealing on the black market. If shots are exchanged, the next day the paper runs a short article telling that a dangerous repeat offender shot at police who had come to arrest him and that the police had killed him.

We have just described the direction-finding method employed at the beginning of 1943. This method, almost completely automated through electronics, was even then highly perfected.

Let's summarize the times required:

Direction-finding of the suspect frequency by the Paris station: instantaneous
Transmission by direct telephone line to position-finding bases: 2 minutes
Direction-finding by the 3 bases: instantaneous
Transmission to the operations room, representation of azimuths on the map, and location of the goniometric triangle: 7 minutes
Transmission of coordinates to the closest mobile bases and dispatch of vehicles: 5 minutes
Total: 14 minutes

If the mobile direction-finding base is in the same city as the clandestine transmitter, the transmission site can be located within a 200-yard radius in less than a quarter of an hour.

In 1943, one could drive through Paris, north to south, in less than 10 minutes. Therefore, if a transmitter is operating for the first time, a direction-finding vehicle can be within sight of the station in less than a half hour after its first signal.

For subsequent transmissions, the Paris central would transmit the frequency directly to the mobile base to reduce the time involved:

Direction-finding of the suspect frequency: instantaneous
Transmission from Paris to the mobile base, visual display of the frequency, dispatch of vehicles: 5 minutes

If the clandestine operator is working inside the city, he is localized within 10 minutes, or 15 minutes after first signal.

If transmissions become regular, the mobile base itself monitors the frequency and direction-finding is reduced to 10 minutes.

It is likely that by the spring of 1944 the Germans were using a completely automated direction-finding system, even in the vehicles. This was achieved by a cathode-ray screen. With this system, it is very likely that the 3-second acknowledgment transmission from the clandestine operator to the home station would have given an azimuth with an approximation of about 5 degrees. This would have enabled localization within a circle of a radius of half a mile.

But no automation, however advanced, would reduce the length of the final closing-in with the field-strength monitor. This might have lasted a long time in a city and thereby risked attracting the attention of the lookout or the guards who surrounded the radio operator whenever feasible.

By establishing 5 minutes as the maximum for a transmission on a given frequency, the S.O.E. gave its radio operators the greatest chance to avoid capture.

Transmission length had certainly not been determined by mere estimation. Secret documents which were distributed to the clandestine operators in 1943, used for this study, prove that the penetration agents of the Intelligence Service and the Service des Renseignements were completely up-to-date on the entire German disposition from December 1942, even knowing the license-plate numbers of the reconnaissance cars.

German detection methods had made decisive progress in 2 years. In 1941 and 1942, the localization of a clandestine station was extremely difficult. It could only be carried out if the operator transmitted on the same days of the week, from the same site, and on the same frequency during several consecutive hours. Direction-finding operations were not yet automatic, and panoramic reception was nonexistent. The scanning of all usable frequencies was necessarily very slow and left substantial gaps.

In addition, during the final approach, each Gestapo agent had to hide a heavy suitcase containing a receiver with a loop aerial under his coat. A Tirolean cap or Basque beret tilting down over his ear just barely hid an

# Position-Finding Technique

## *The First Phase of Position-Finding*

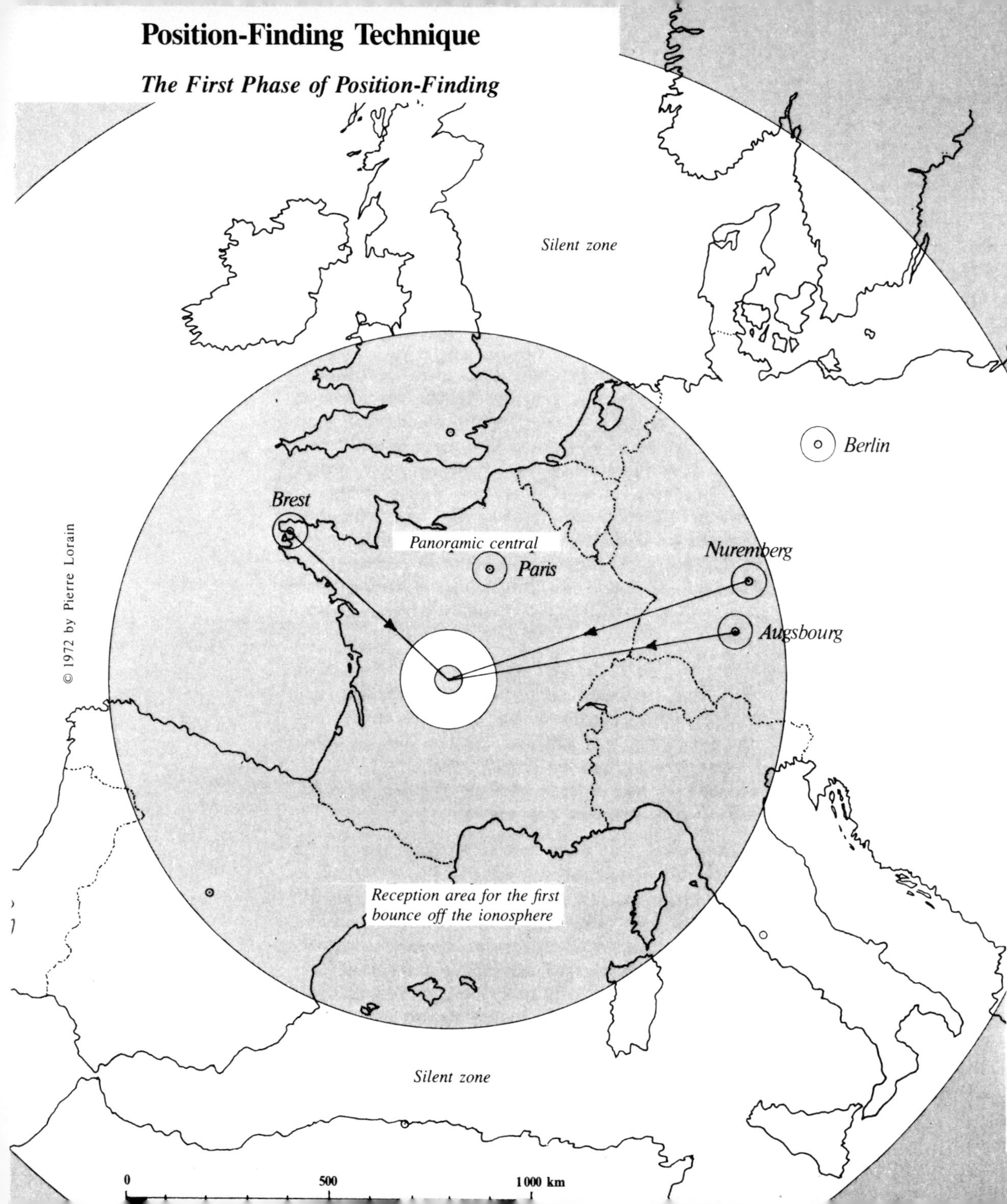




The gray concentric rings represent the areas normally covered during the day by a transmitter of 4 watts, transmitting on a frequency of 7 megacycles.

The first phase of position-finding delimits the clandestine station within a triangle 10 miles on each side.

Alerted by the panoramic monitoring central at Paris, the far-off bases of Brest, Nuremberg, and Augsburg seek the transmitter's position.

Note that the base at Berlin, located in the silent zone at this hour of the day, is unable to pick up the clandestine station.

The first phase of position-finding is unavoidable in view of the automatic and systematic monitoring of all frequencies. This provides an initial, imprecise localization of a presumed center of resistance, but it is not dangerous, in itself, for the operator.

NOTE: This map of theoretical average wave propagation was established through experiments with an S.O.E. set, Model A Mark III, for a frequency of 7,000 kilocycles, operated about 10 A.M. in April, from the Creuse department in central France.

## *Second, Third, and Fourth Phases of Direction-Finding*

SECOND PHASE OF DIRECTION-FINDING

*The second direction-finding phase locates the clandestine station within a triangle measuring ½ to 1 mile along the sides. Direction-finding is carried out, within the range of the ground wave, by a stationary regional base (in this case, Clermont-Ferrand) and 2 cars seeking the direction from which the signal is coming from their locations at the apexes of the triangle determined by the first phase.*

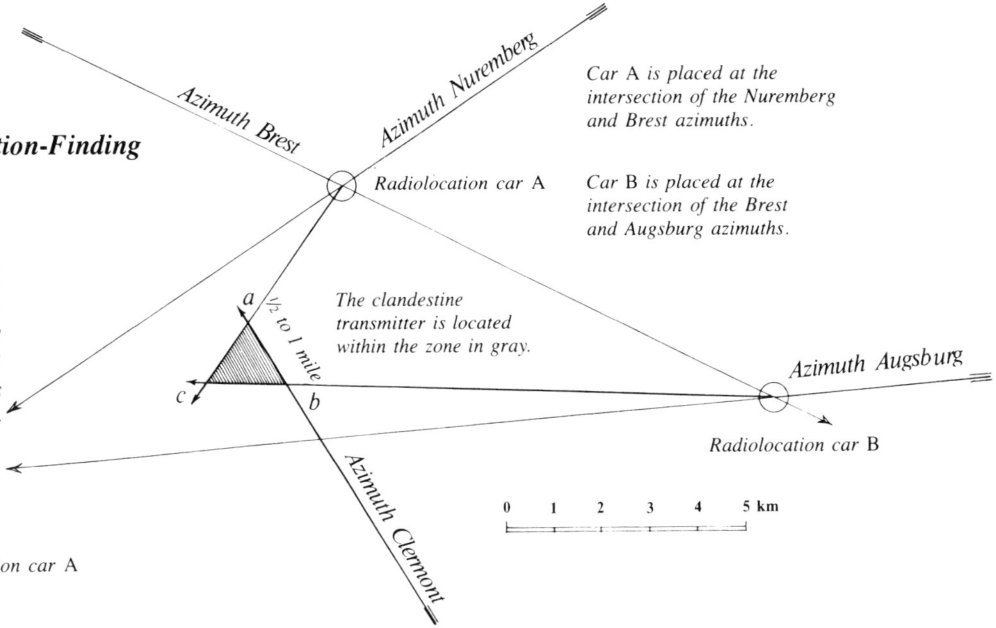


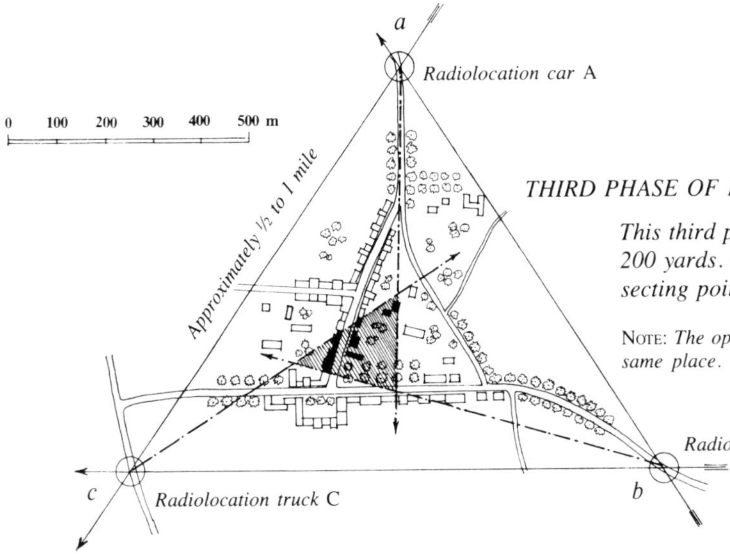


THIRD PHASE OF DIRECTION-FINDING

*This third phase locates the clandestine station within a radius of approximately 200 yards. Direction-finding is carried out by* A, B, *and* C *vehicles at the intersecting points of the triangle defined by the second position-finding phase.*

NOTE: *The operator is not yet in immediate danger, but he should move and never return to the same place.*

FOURTH AND FINAL PHASE OF DIRECTION-FINDING

*The final phase of approach is made with the field monitor in the 200-yard zone defined by the third phase. Pedestrians equipped with miniature field-strength meters and walking down the main street of the area discover the transmitter inside the building indicated in gray. The operator is in immediate danger. If the troops have set up barricades, he is practically lost. Toward the end of 1943, the operating schedules handed over to each operator were, in theory, to prevent the fourth phase of position-finding from taking place.*

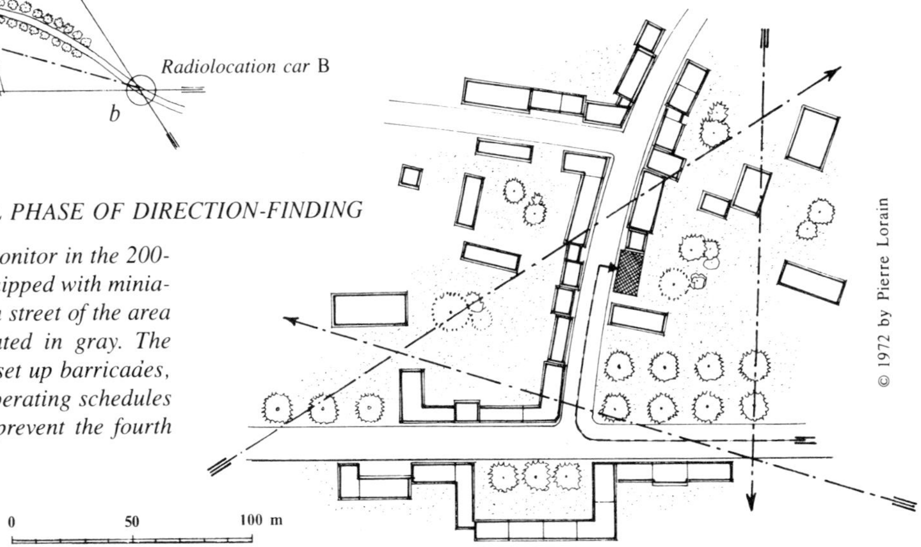

earphone. Their general posture aroused the curiosity of even the most naïve of passersby.

The arrest of a radio operator thus often required long months of continual surveillance. The operation was complicated by the fact that if a clandestine operator was spotted in the unoccupied zone of France (controlled by Vichy), the Germans could only signal the suspect frequency to the French radio control group at Hauterive near Vichy. The latter promised to look into the matter, but secretly warned the clandestine station to move as quickly as possible, and then supplied the Germans with an almost completely false position.

## THE EVOLUTION OF CLANDESTINE COMMUNICATIONS FROM 1941 TO 1944

This evolution followed closely behind the evolution of German direction-finding technique.

During the first half of the war, the clandestine operator carried a heavy suitcase containing a radio set and quartz crystals around with him. He spent several weeks, or even several months, set up at a safe address he had been informed about before he left. Procedure required him to establish contact with the home station when he had a telegram to transmit. If reception was poor, the base might ask him to repeat several groups or the entire message. After acknowledgment, the home station transmitted any messages it might have for the network.

The operator could therefore have spent several hours per day next to his set in 2-way communication. In addition, he had to encipher or decipher messages whenever there were communications. Finally, for lack of another address, he often slept where his set was.

When someone told him about new, friendly premises, he left with his suitcase, generally traveling by train toward a destination which could be very far from the previous site.

During this period, almost all clandestine broadcasts took place in the unoccupied zone for reasons we have already considered.

But the situation changed on November 11, 1942, 3 days after the Allied invasion of French North Africa, when the Germans, in Operation ATTILA, invaded the unoccupied zone, quickly and decisively implanting their Gestapo and regional direction-finding stations everywhere. Soon the weight and bulk of the English sets diminished, the number of available crystals increased, and safe addresses became more numerous. The radio operator still moved around with his suitcase, but he now worked according to a well-established plan of operation on 3 or 4 different frequencies. He often traveled by bicycle because the new, smaller suitcases could be securely tied down to the luggage rack with two rubber straps. The procedure still consisted of conversing with the home station, but the call letters changed at every session according to a microfilmed key.

Finally, in 1944, there arrived the height of automation in direction-finding technique. But safe addresses were numerous and enough sets were parachuted in. Furthermore, these sets had diminished in weight and bulk compared to the older ones. Many radio operators had 5 frequencies and 5 transmitters, hidden in 5 discreet locations about 5 miles apart from one another.

Procedure had changed radically. Except for emergency messages, there were no longer any simultaneous exchanges between the base and the clandestine station. Transmission and reception were split into 2 hermetic compartments and were no longer carried out on the same premises. At night in his lodgings, the radio operator listened for the telegrams which were intended for him; it was also there that he performed his enciphering and deciphering. He alone knew about this address. He returned to his different transmission sites by bicycle and carrying nothing. He used these sites in rotation and moved on at the end of a few weeks so as not to exceed 3 transmissions on the same frequency at each given place.

The operator never carried his equipment himself, and the crystals traveled separately. An armed courier, who served him as bodyguard, followed him like a shadow. Finally, he used a cryptographic system that was mathematically unsolvable by any current or future equipment.

In this way, security was at a maximum. The number of networks transmitting simultaneously saturated the German services, which no longer had enough personnel to cope. One or 2 radio operators were still captured as a result of radiogoniometry but only because the Germans concentrated on one operator at a time while disregarding the others.

Finally, after the landing in Normandy, when entire sections of France were completely liberated, the secret and clandestine nature of the S.O.E. stations faded away. It was completely safe for the operators to carry out their work in the same manner as classic military stations. In other words, they could stay at the same place, set up good outdoor aerials, and work the entire day if necessary. Protected by powerfully armed units against any unlikely sudden raids by German police, they were invulnerable. Some even had their officers' uniforms dropped by parachute!

Within weeks, the character and the role of these stations lost their *raison d'être*. At the end of September 1944, the S.O.E. recalled all its missions, packed its bags, and left France to concentrate all of its efforts in Germany and the Far East.

## THE FIRST SETS

The British secret services used many different kinds of transceivers during the war. These underwent many changes. Some were only prototypes employed for a single operation. Others had ephemeral life-spans and were promptly abandoned after revealing incurable defects during operations.

To describe them all would be tedious. Besides, this survey is limited to a detailed study of those sets which, because of their importance or qualities, marked these 4 years with rapid progress and constituted the larger part of sets at the disposal of I.S. and S.O.E. agents.

Their evolution between 1941 and 1945 showed a tendency toward a continual reduction in weight and bulk and, at the same time, a growing complexity of circuitry techniques. Consequently, during the war, compactness increased. A partial and then a total miniaturization of the components followed in 1944 and 1945. The energy beamed by the clandestine transmitter likewise diminished as the home stations equipped themselves with high-sensitivity receivers.

Therefore, a transceiver weighing more than 45 pounds in 1941 was replaced at the beginning of 1944 by a set which did not exceed 10 pounds. In 1945 the weight fell below 5 pounds. Receivers were the great beneficiaries of technical progress: they went from mediocre to very good.

The sets are shown in chronological order. This is not to say that the newer model pushed the earlier model out of use. Each set was too valuable and too much needed not to be employed to its fullest extent. This was why, at the end of hostilities in France in 1944, the I.S. Paris station still housed a primitive and outmoded Mark XV receiver.

The Mark XV transceiver is, without a doubt, the oldest of its kind. Its technical design goes back to the last years before hostilities broke out. It was built around 1938. Although it was carefully worked out, it was somewhat amateurish. The varnished wooden boxes were antiquated and were difficult to carry. The set was assembled with commercial components used by amateur radio operators around 1937.

Two suitcases were needed to carry the complete set. The transmitter had good power and permitted communications over great distances because of the wide range of frequencies it covered. It was equipped with an excellent quiet micro key and permitted remote control by means of a separate key connected by a long cable.

The receiver fell short of the mark on account of its extremely outmoded technical makeup, which gave it poor selectivity and very delicate tuning. Its stability was mediocre. However, it should be noted that this set, designed before the S.O.E. existed, was destined for the MI-6 intelligence networks. This justified special attention to the transmitter and called only for a receiver good enough to pick up short messages from the home station.

The Mark XV was the set used for the first missions in France. An updated version, which probably appeared around the beginning of hostilities, was the Paraset.

## Mark XV Transmitter

*Half size*

Standard Features

*Frequencies covered: 3.5–16 megacycles in 3 wave bands*
*Power output: 15–20 watts*
*Two American metal tubes (6F6 crystal-controlled pilot; 6L6 amplifier)*
*Built-in microkey. The combined power pack for house current and 6-volt battery (not shown) are contained in a heavy metal box.*
*The weight of the instrument is about 45 pounds. One is on display at the Army Museum in Paris.*

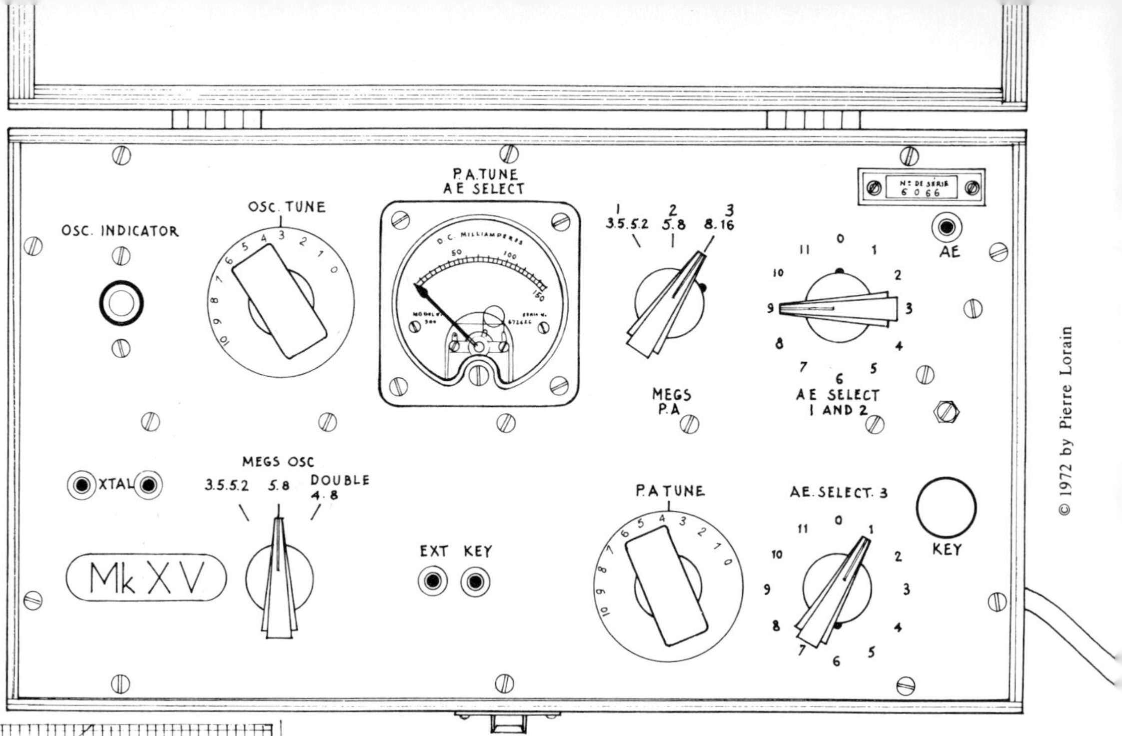




## Mark XV Receiver

*Half size*

Standard Features

*Straight regenerative receiver*
*Three American metal tubes*
*Frequencies covered: 3–13 megacycles in 1 wave band*

*Note the fine tuning dial which permits easy tuning with a frequency range of 200 kilocycles on either side of the frequency displayed on the main dial; the calibration curve, attached to the inside of the lid, and the reaction and sensitivity controls for Morse reception. Transmitter and receiver are contained in 2 varnished laminated boxes. The brown bakelite front panels are metal-plated inside.*

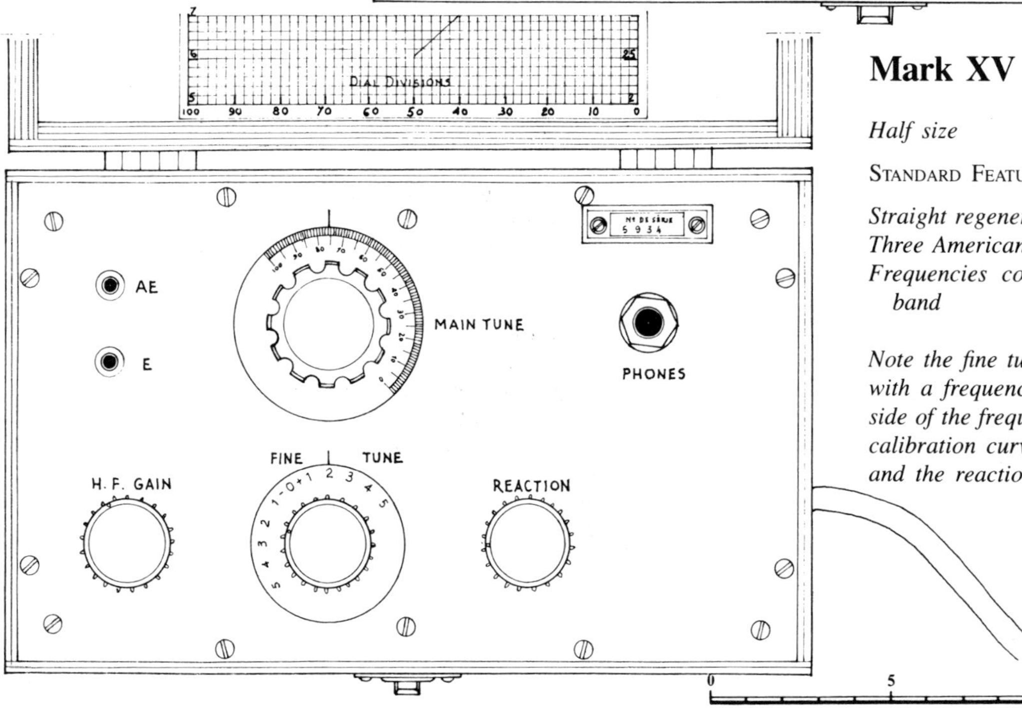


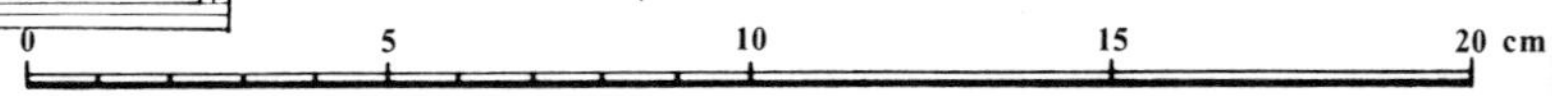

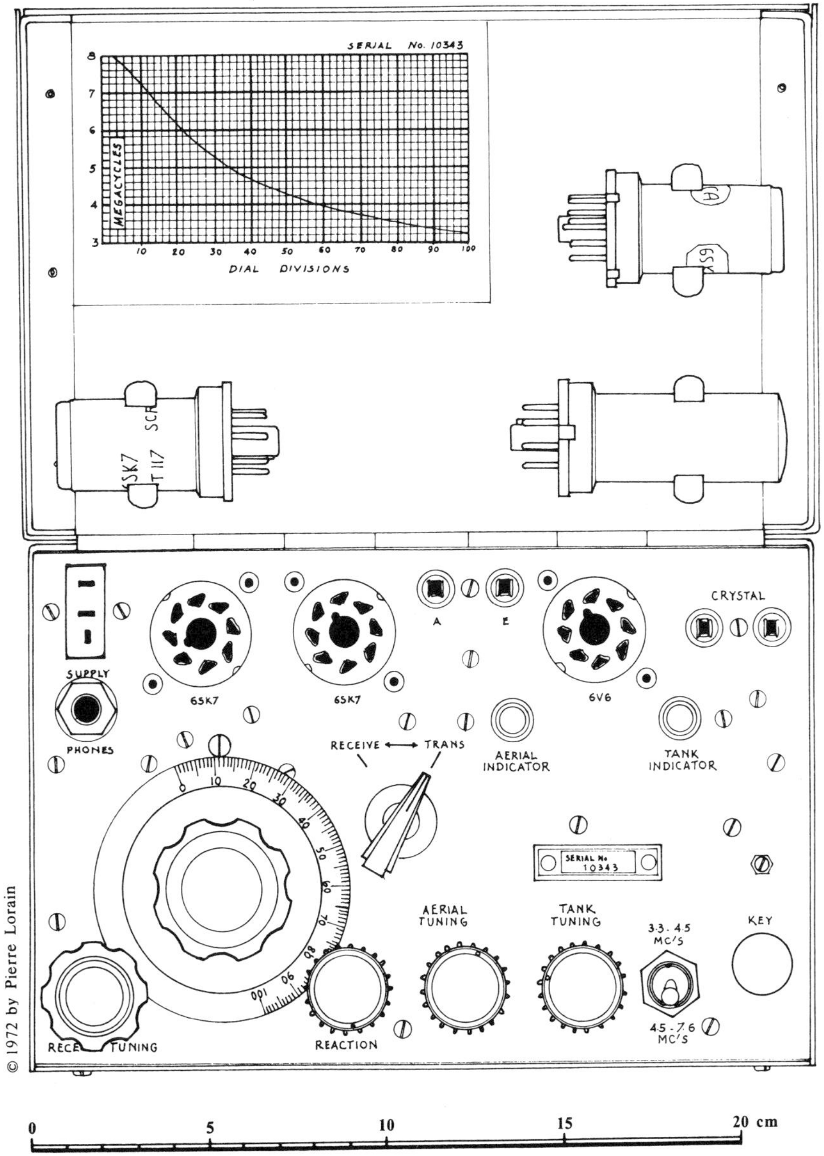




# Paraset Transceiver

*Half size*

STANDARD FEATURES

*Frequencies covered: 3–7.6 megacycles in 1 wave band for the receiver and 2 for the transmitter, 3.3–4.5 megacycles and 4.5–7.6 megacycles*
*Power output: 4–5 watts; 3 American metal tubes*
*Single tube transmitter: 6V6 crystal-controlled pilot. Built-in micro key*
*Straight regenerative receiver: 2 6SK7 tubes*
*Cadmium steel box, silver finish. Dimensions: $8\frac{1}{2} \times 4\frac{1}{4}$ inches*
*Weight: About $3\frac{1}{4}$ pounds*
*(The combination power pack for AC and 6 volts battery operation is not shown. It is housed in a similar box with an average weight of $6\frac{1}{2}$ pounds.)*

***View of Opened Box***

*Under the lid there is a calibration curve for the receiver and 3 metal tubes held by clamps.*

*On the front panel, left to right and top to bottom: jacks for supply cord and headset, sockets for the receiver's 2 tubes, sockets for aerial and ground wires, socket for the transmitter's tube, and crystal holder.*

*Under the 6V6 socket, 2 light bulbs for transmitter tuning.*

*Below, the receiver dial with vernier for fine tuning, the regeneration control, the 2 transmitter tuning knobs, the wave-band toggle switch and the built-in key.*

*Switching from "Transmit" to "Receive" is carried out with the arrowhead knob in the middle of the panel.*

### *View of Paraset in Operation*

*With the box open, the supply cord is connected and the headset, aerial and ground wires, tubes, and crystal are plugged in. The operator is transmitting a message.*

*The initial tuning is extremely easy:*

*After plugging in the valves and headset, selecting the right crystal, and connecting the power pack, the operator tunes the receiver to the home station frequency, using the calibration curve pasted under the lid.*

*He then switches to "transmit," depresses the key, and simultaneously turns the tank tuning and aerial tuning knobs until both bulbs glow at their brightest. The set is then properly tuned and ready for operation.*

*For maximum sensitivity and selectivity during reception, the operator has to turn the regeneration control up until a hissing sound can be heard through the headset.*

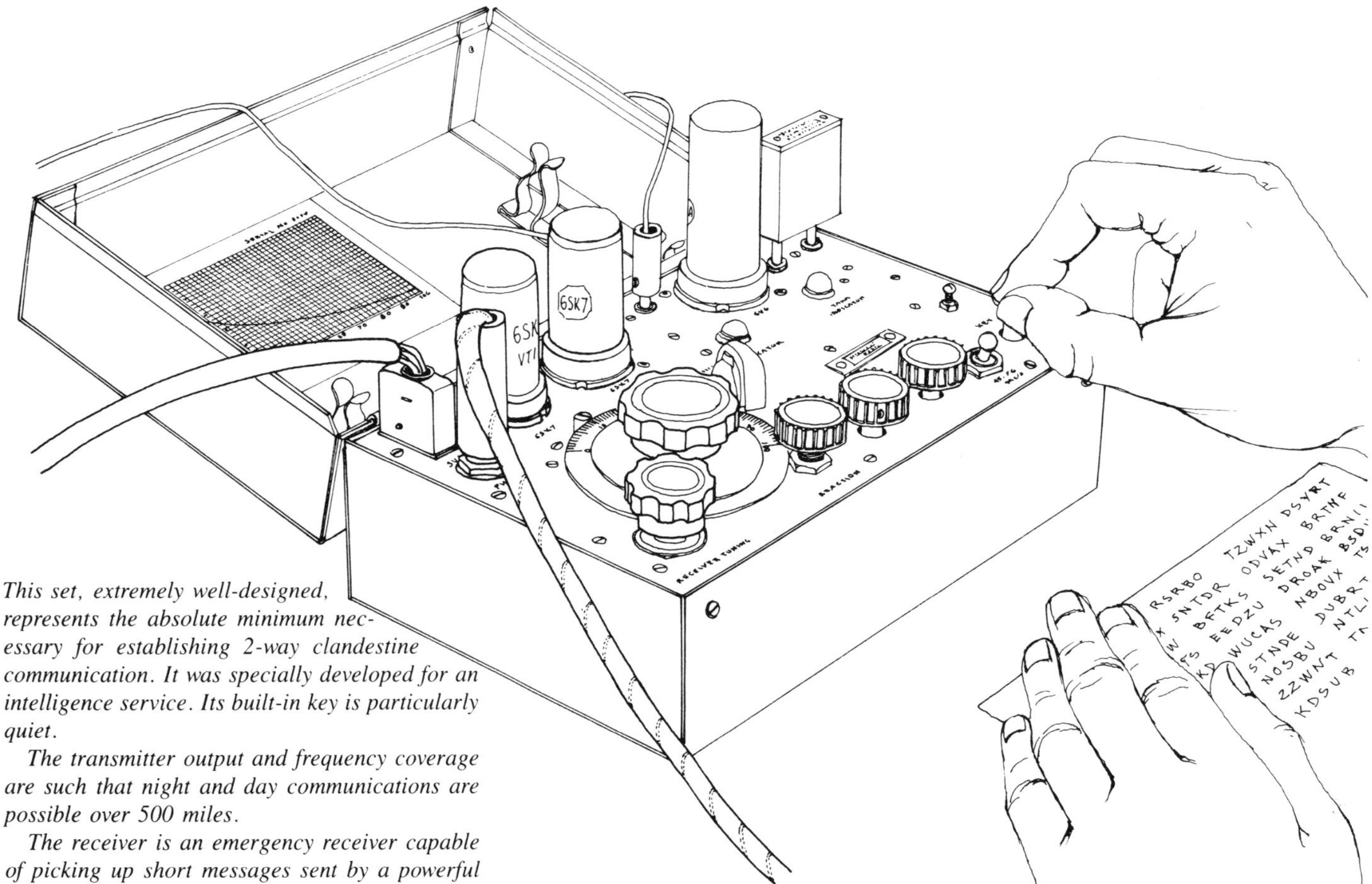




*This set, extremely well-designed, represents the absolute minimum necessary for establishing 2-way clandestine communication. It was specially developed for an intelligence service. Its built-in key is particularly quiet.*

*The transmitter output and frequency coverage are such that night and day communications are possible over 500 miles.*

*The receiver is an emergency receiver capable of picking up short messages sent by a powerful home station.*

# THE POLISH SETS

### The B.P. 3 and A.P. 4 Transceivers

In 1941, at Letchworth, 30 miles north of London, a group of Polish civilian refugees set up a small electronic instruments factory. Very discreet, it never attracted any attention. At first sight, one would have thought it a crafts shop, of which there were so many. In reality, these colorless civilians without pasts were remarkably expert in telecommunications. They built, in the greatest secrecy, transceivers for the I.S. networks. The advanced technology of their devices pushed all other existing devices down to the status of museum pieces.

The Polish series was certainly one of the most brilliantly designed in the domain of clandestine transmissions. The high quality of finishing afforded the smallest details reveals both the German school, of which the Poles had been students, and the natural desire of the weak to assert themselves in the face of the British secret services and to prove that Poles were capable of supplying the best, most technologically advanced sets.

Between 1941 and 1944, about 10 models were manufactured in substantial quantities. We will describe the most famous of these: the B.P. 3 and the A.P. 4. The initial models, A1 and A2, left the famous Paraset transceiver far behind. It was not until 1943 and 1944 that the British specialists produced sets of equal and perhaps superior quality in the Type 3 sets Mark II and Mark III.

The Intelligence Service grabbed up the B.P. 3 and A.P. 4 sets to equip their secret bases in Paris, Lyons, Bordeaux, and Rheims. A total of 24 Polish sets ensured constant communications between MI-6 agents and their British home station.

For many months, the Polish sets were not only the best, but also the most compact. The A.P. 4 could be considered the model of the genre. Equipped with some miniaturized components, an excellent receiver housed in a very small box, a transmitter of adequate power, and using a 110/120-volt external power supply, this set was certainly one of the best for a clandestine city network. An additional advantage was that it could be hidden, for example, under parquet boards, and operated by remote control (the socket marked *K* was reserved for that).

Transmitter tuning was accomplished in a few seconds. The operator set the *S-N-O* switch on the *S* position ("tune"), depressed the key, and turned the arrow-marked knob until neon tube *Z* lit up. The operator then switched to *N* (*Nadajnik*, for "transmit") and chose whichever of the 12 positions on the upper arrowhead knob most lit up the tuning bulb underneath. The transmitter was then ready to operate.

By turning the switch to *O* (*Odbiornik*, for "receive"), the operator turned on the receiver. On the control board, the regeneration control (*Reakcja*) in the detection circuit considerably increased the sensitivity and selectivity of the set while enabling it to receive telegraphic signals.

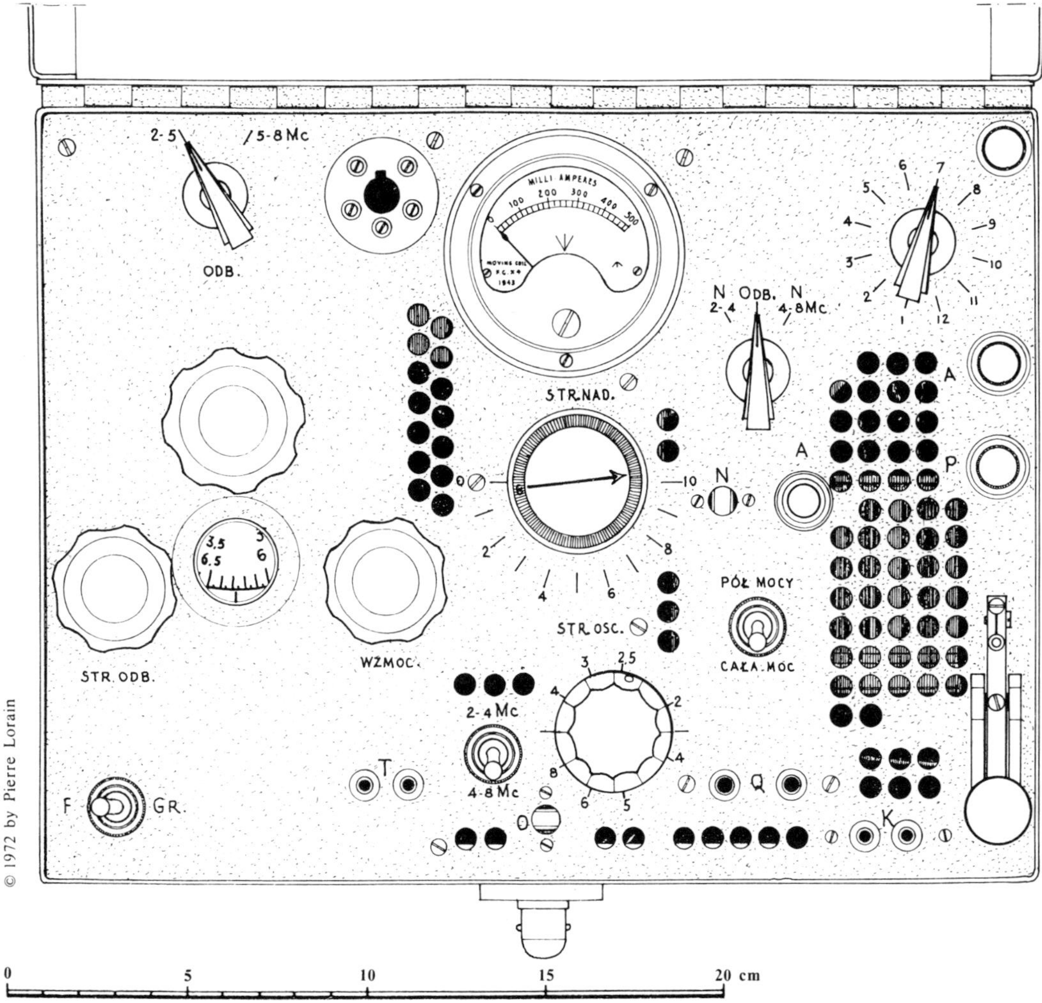




# B.P. 3 Transceiver

*Half size*

STANDARD FEATURES

*Coverage: 2–8 mc/s in 2 wave bands*
*Power output: 30 watts average*
*2-tube transmitter, 1 is dual-purpose (crystal-controlled oscillator and amplifiers)*
*Superheterodyne receiver with 4 American metal tubes*
*Steel box with black wrinkle finish, inscribed white letters*
*Dimensions: 11 × 8½ × 4 inches*
*Built-in key. The combination 12-volt auto battery and power supply pack is not shown. It is contained in a box, the same size as above, and weighs about 15½ pounds.*

***View of Opened Box***

*Directions for use, a diagram of the set, and a list of recommended types of aerials are all pasted under the lid. This could be written in either Polish or English.*

*On the front panel, the area to the left of the milliammeter is for the receiver. Present are the wave switch for 2–5 megacycles and 5–8 megacycles, the 5-pin supply-cord socket, the frequency control dial, its fine tuning control (STR ODB), volume control (WZMOC), the telephone/telegraph switch, and the headset socket.*

*The rest of the panel is reserved for transmitter controls. These are relatively numerous since the transmitter is powerful for its size.*

*The neon tubes marked* N *and* O *and bulb* A *show that the tuning is correct when they light up as brightly as possible. The 3 terminals along the righthand side are ground and aerial terminals.*

*On the lower part, the* Q *socket is for the quartz crystal, and the* K *socket, for a detached key.*

### *View of B.P. 3 in Operation*

*The box is open and the power supply plug and quartz crystal are in place. The radio operator is transmitting. Preliminary tuning is effected as follows:*

*After installing the appropriate quartz crystal and switching on the desired wave band, the operator pushes the switch marked* POŁ MOCY/CAŁA MOC *(half power/full power) down to "half power" and depresses the key while turning the knob marked* STR.OSC *until neon tube* O *lights up. The milliammeter should then indicate about 110 milliamperes.*

*The operator then switches to "full power." He adjusts the dial marked* STR.NAD *and the 12-position arrow knob until the neon tube* N *and the* A *bulb light up. The milliammeter should then indicate 210 milliamperes. The transmitter is ready to operate.*

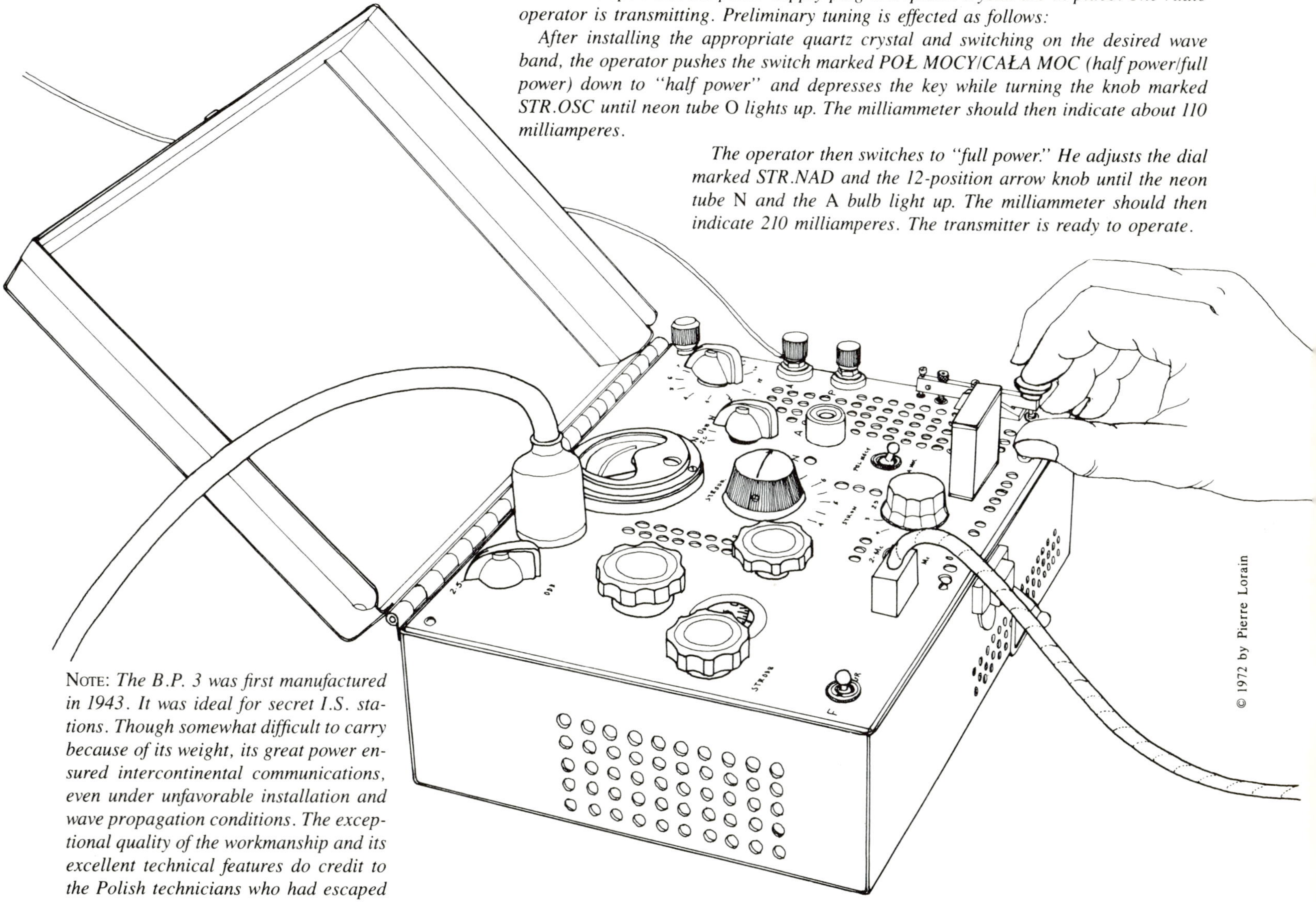



NOTE: *The B.P. 3 was first manufactured in 1943. It was ideal for secret I.S. stations. Though somewhat difficult to carry because of its weight, its great power ensured intercontinental communications, even under unfavorable installation and wave propagation conditions. The exceptional quality of the workmanship and its excellent technical features do credit to the Polish technicians who had escaped to London.*

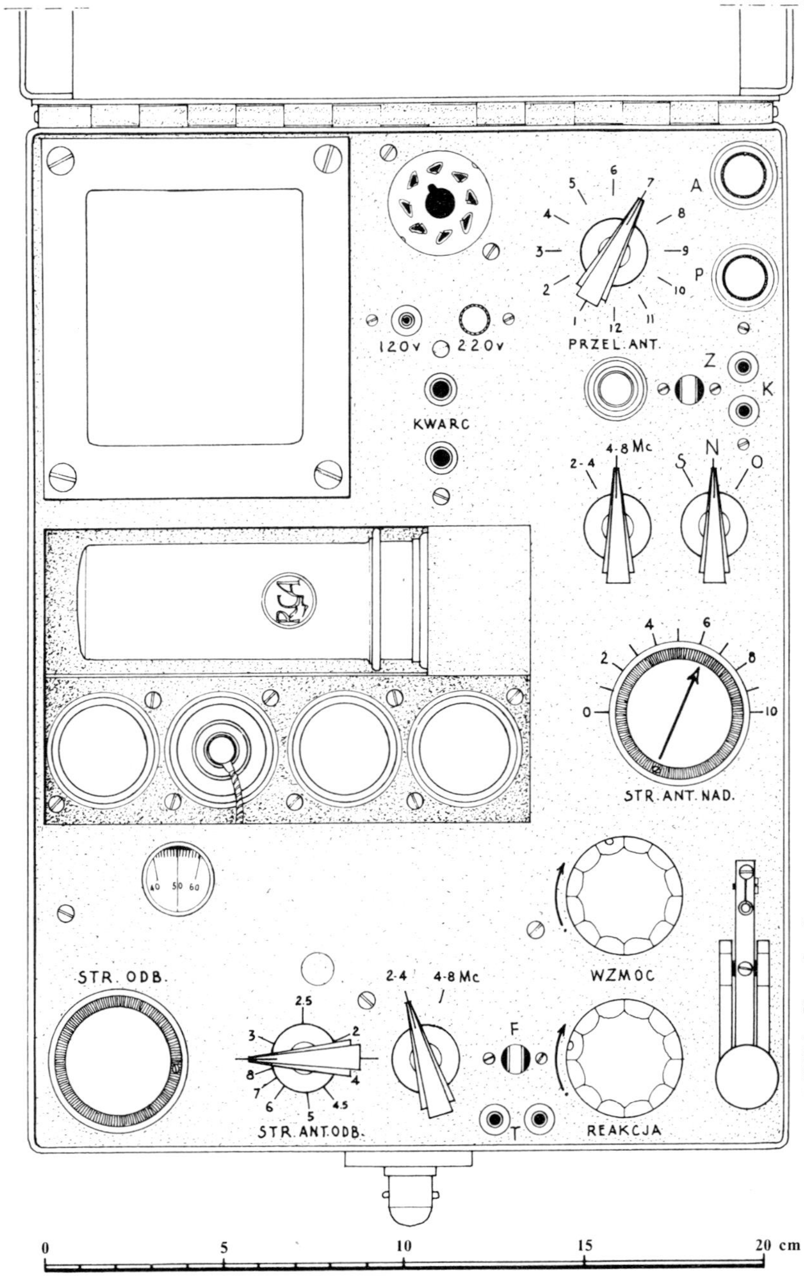
RCA
120v
220v
KWARC
PRZEL. ANT.
A
P
Z
K
2-4
4-8 Mc
S
N
O
STR. ANT. NAD.
WZMOC
STR. ODB.
2.4
4-8 Mc
STR. ANT. ODB.
F
T
REAKCJA
0
5
10
15
20 cm

# A.P. 4 Transceiver

*Half size*

STANDARD FEATURES

*Coverage: 2–8 megacycles in two wave bands*
*Power output: 8 watts average*
*Transmitter with 1 crystal-controlled oscillator*
*Superheterodyne receiver with 3 metal American tubes*
*Built-in A.C. power pack 120/220 volts in a single, steel-plated box, black wrinkle finish.*
*Dimensions: 11 × 8½ × 4 inches*
*Weight: About 9 pounds*

***View of Open Box***

*Inside the lid are directions for use, a circuit diagram, and 2 charts for the conversion of dial graduations into kilocycles.*

*The upper half of the panel is set aside for the power supply and the transmitter, including the transformer and the socket for the supply cord. Underneath this socket, one can see the voltage selector and the crystal holder.*

*The transmitter is adjusted until neon lamp* Z *and the incandescent bulb light up to their brightest. The switch for tuning, transmitting, and receiving is marked* S-N-O.

*In the recessed area of the box, 5 tubes are accessible: the horizontal transmitter tube, the rectifier tube, and the 3 side-by-side vertical receiver tubes.*

*The lower half houses the receiver. Neon lamp* F *above headset sockets* T *lights up when the receiver is tuned to the transmitter's frequency.*

# THE S.O.E. SETS

## A Mark II* Suitcase Transceiver

Since the S.O.E. was authorized to build its own sets beginning in June 1942, the A Mark II* model, which came out early in October of that year, could be considered one of the S.O.E.'s very first sets.

It was preceded by the 21 Mark I (August 1942), which had obsolete, bulky, fragile tubes. The innovative A Mark II* adopted a new type of tube from the United States. These excellent "Loctal" tubes were strong, modern, and compact. They were a step toward miniaturization and would be used from then on until the end of hostilities for all British secret service sets.

The A Mark II* was made up of 3 metal compartments of identical size. The complete set was somewhat bulkier and heavier than the Polish A.P. 4. Its receiver had the same characteristics as the A.P. 4, but its transmitter was weaker. Finally, although they had good quality components, the A Mark II* sets showed all the symptoms of hasty, homemade construction. However, its users appreciated the fact that it could be easily carried in 3 separate and discreet small parcels.

STANDARD FEATURES

*Coverage: 3–9 megacycles in 2 wave bands; "Blue" wave band, 3–4.5 megacycles, and "Red" wave band, 6–9 megacycles*
*Power output: 5 watts average*
*Transmitter with single crystal-controlled oscillator*
*3-tube superheterodyne regenerative receiver. External power of 100–250 volts or automobile battery of 6 volts (a 6-volt nonsynchronous vibrator is inside the combination power pack).*
*3 metal boxes, black enamel finish on front facing. White inscriptions, painted by hand or stenciled.*
*Dimensions of each box: about 11 × 4 × 3 inches*

*A small accessories box containing the headset, the key, an extra crystal, 2 replacement tubes for the receiver (a replacement tube for the transmitter is found underneath the lid of the box), 8 fuses, a screwdriver, a neon lamp, and a cord to the house current. Coiled around the box are about 60 feet of antenna wire wrapped in green plastic and 10 feet of ground wire in black plastic.*

*The total weight of the suitcase is approximately 20 pounds.*

***View of Open Suitcase***

*The left box houses the receiver: a 2-speed tuning dial in the upper right, wave band switch, regeneration, and volume controls. In the upper left-hand corner, a neon lamp lights up when the receiver is tuned to the transmission frequency.*

*The middle box contains the transmitter. From top to bottom are the send/receive switch, the aerial terminal, the milliammeter, and 2 tuning knobs. Below are the crystal holder, the key socket, and connecting cables to the receiver and power pack.*

*The right-hand box is reserved for the power supply. Halfway down are the voltage selection bridge and a clamp for the key. At the bottom are the power switch, the supply cord and its socket, and a fuse.*

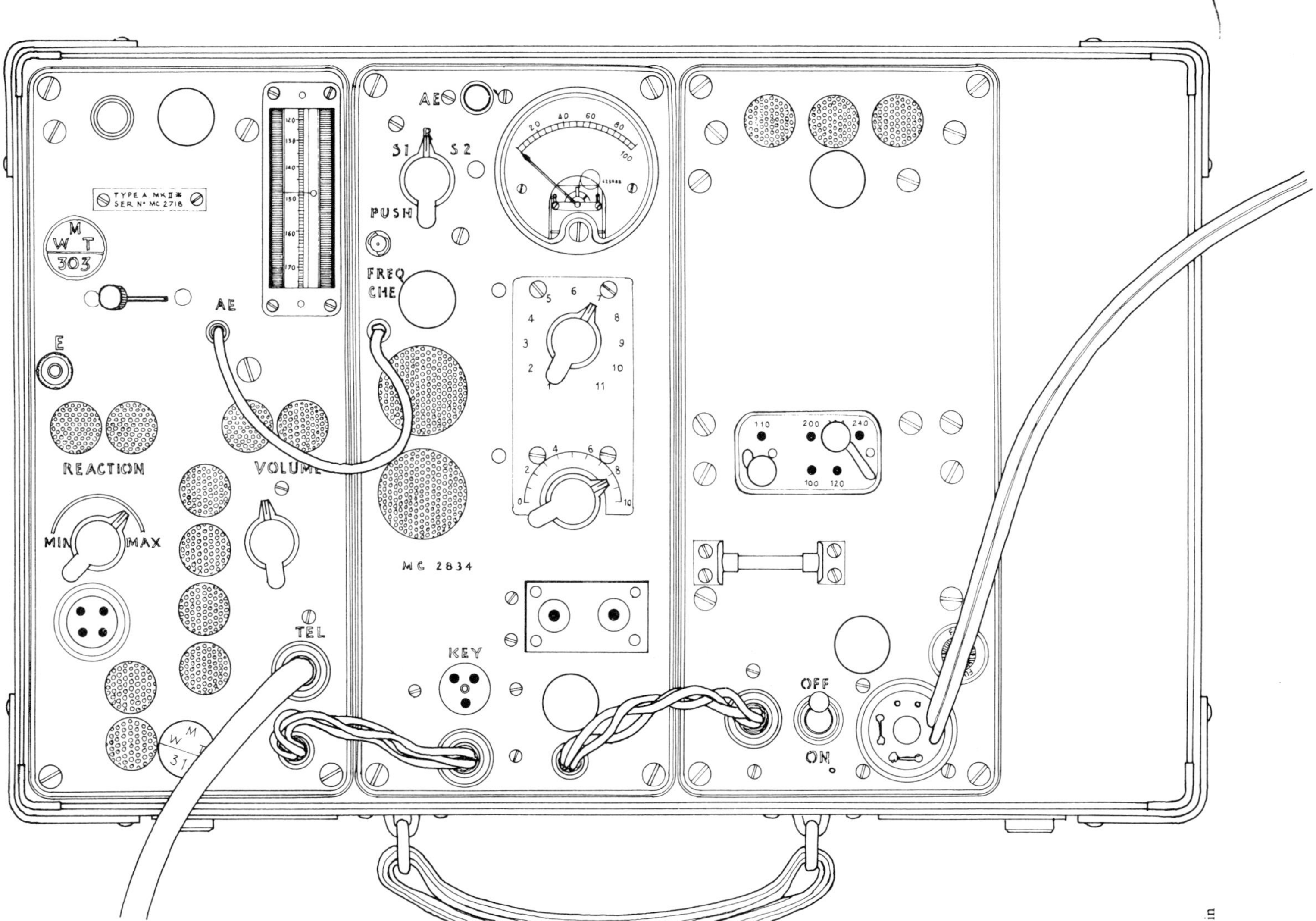


*Half size*

0 5 10 15 20 cm

# 3 Mark II ("B2") Suitcase Transceiver

Standard Features

*Coverage: 3–15.5 megacycles in 3 wave bands for the receiver, and 3–16 megacycles on 4 interchangeable coils for the transmitter*
*Power output: 30 watts minimum*
*2-tube transmitter: quartz crystal-controlled oscillator and power amplifier (EL32 and 6L6)*
*Superheterodyne receiver with 4 "lock-in" multiple tubes*
*Power supply of 97–250 volts from house current and 6 volts from auto battery, with vibrators in the same compartment*

The 3 Mark II set, commonly called the B2, was, like the A Mark III, one of the most highly perfected clandestine sets produced during the war. From 1943 until 1945, and probably beyond, it was used and appreciated more than any other medium-power set.

It made possible easy long-distance communication. Its receiver was particularly selective and effectively rejected interfering signals.

It was usually camouflaged by a small suitcase divided into 4 compartments. The left-hand one housed an accessories box, which was covered by a lid onto which a key could be screwed and contained 60 feet of aerial wire, 10 feet of ground wire, supply and battery cords, a lamp-socket outlet (for a plug to get current from a lamp socket), the key, the headset, 4 extra tubes, 4 transmission coils, fuses, and a screwdriver.

The 2 middle compartments housed the following: in the upper half, a transmitter—dimensions, 9 by 7 by 5 inches, weight, 7 pounds; in the lower half, a receiver—dimensions, 9 by 4½ by 5 inches, weight, 7 pounds. The right-hand compartment housed the combined house current–battery power supply—dimensions, 10 by 4 by 5 inches, weight, 12½ pounds.

The weight of the completely equipped suitcase was 32 pounds.

The 4 metal cabinets had a black wrinkle or vermiculated finish. The inscriptions were photoengraved on small black oxidized brass plates. The 3 Mark II was not quite a member of the partially miniaturized generation of sets. However, it was extremely compact.

On the front transmitter panel were the milliammeter and its 6-position selector, which permitted instant monitoring of the principal circuits. The 3 knobs surrounding the quartz crystal holder permitted a crystal of 4 megacycles to transmit on 4, 8, 12, or 16 megacycles, for example.

The 6-pin socket was the interchangeable coil holder. The 2 large graduated knobs controlled the amplifier tube anode circuit and the aerial circuit tuning. They could easily adapt to the most nonsensical aerials, such as clothes lines, zinc rainpipes, metal chimneys, or even metal springs from a mattress.

The fine tuning knob marked B.F.O. could shift about 3 kilocycles from one side of the frequency to the other in the event of interference. The volume knob adjusted receiver sensitivity and desensitized the receiver if the signals from the home station were too powerful. On later models, a magnifying lens protected the graduated frequency dial, which was controlled by a 2-speed tuning circuit.

The power supply cabinet had the voltage adjustment plugs and a plug marked "Mains/Battery." If German interception services cut the "mains" house current to a group of buildings, simply reversing the fitting instantly started the vibrator for battery power. The 1- or 2-second interruption of transmission was negligible and direction-finding networks would not think that the clandestine transmitter was in the block of buildings where they had just cut off the electricity.

This was one of the advantages of the B2 in city operations. S.O.E. operators were strongly advised to plug in their battery whenever possible.

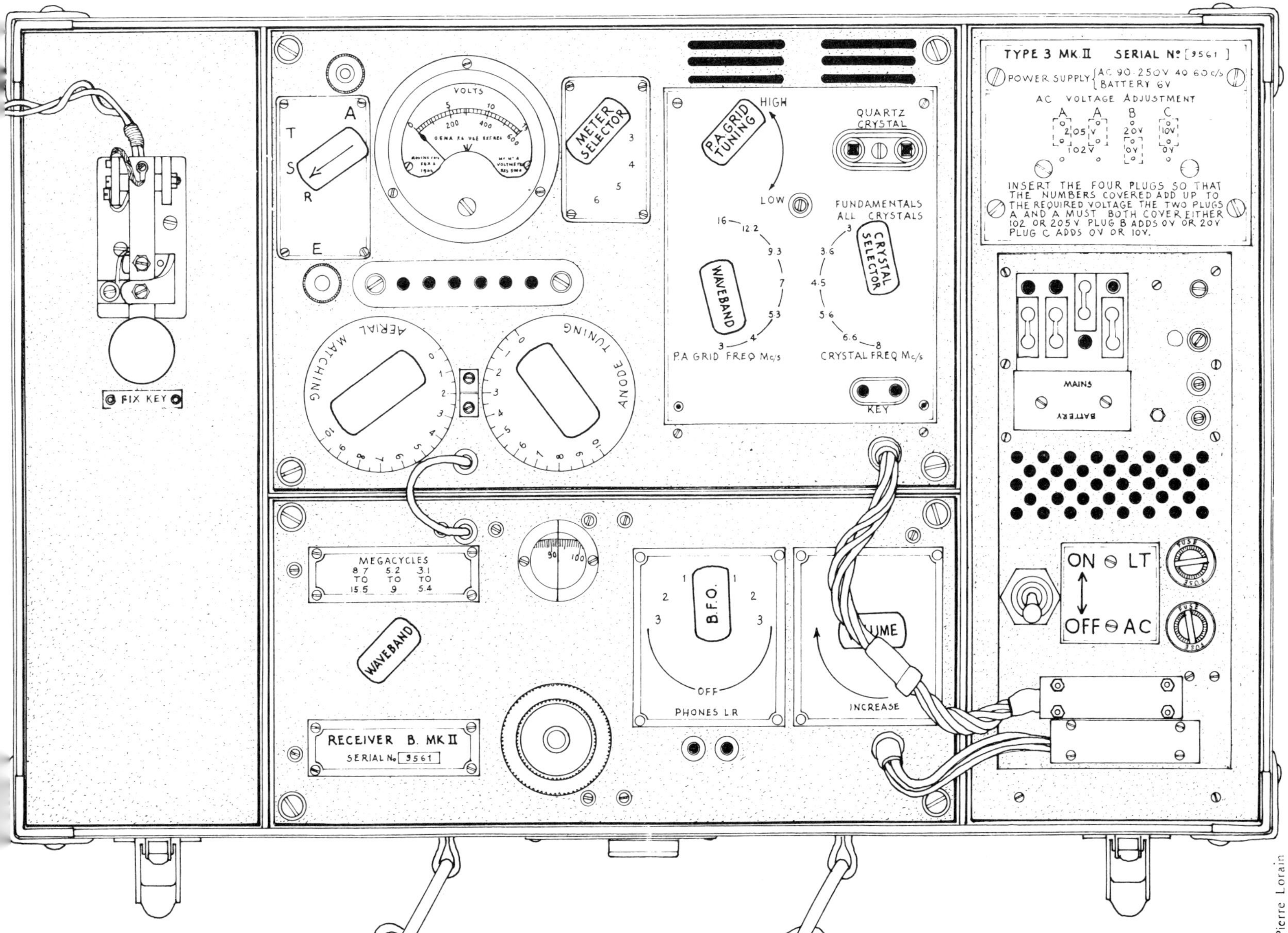


*Half size*

0 5 10 15 20 cm

## A Mark III Suitcase Transceiver

The latest of the S.O.E. sets, the A Mark III, is still admired by today's experts, as much for its design and workmanship as for its technical performance. It easily handled regular communications over more than 500 miles; its receiver was particularly sensitive and stable.

The A Mark III was the smallest transceiver conceived during World War II. It was completely miniaturized, except for the "lock-in" American tubes. It was the last step before the creation of sets with tubes the size of peanuts (transistors would not appear until several years after the end of hostilities). The new radio procedure, separating transmission and reception in time and space, permitted the creation of these miniaturized sets, which were used in small quantities during the very last part of the war and which will be depicted at the end of this chapter.

The A Mark III was dropped by parachute in 2 special airtight containers called Container C and Container D. C weighed 13½ pounds and held the set itself, the headset, and the operating instructions. D weighed 11 pounds and housed the accessories box and the separate power supply for the 6-volt battery.

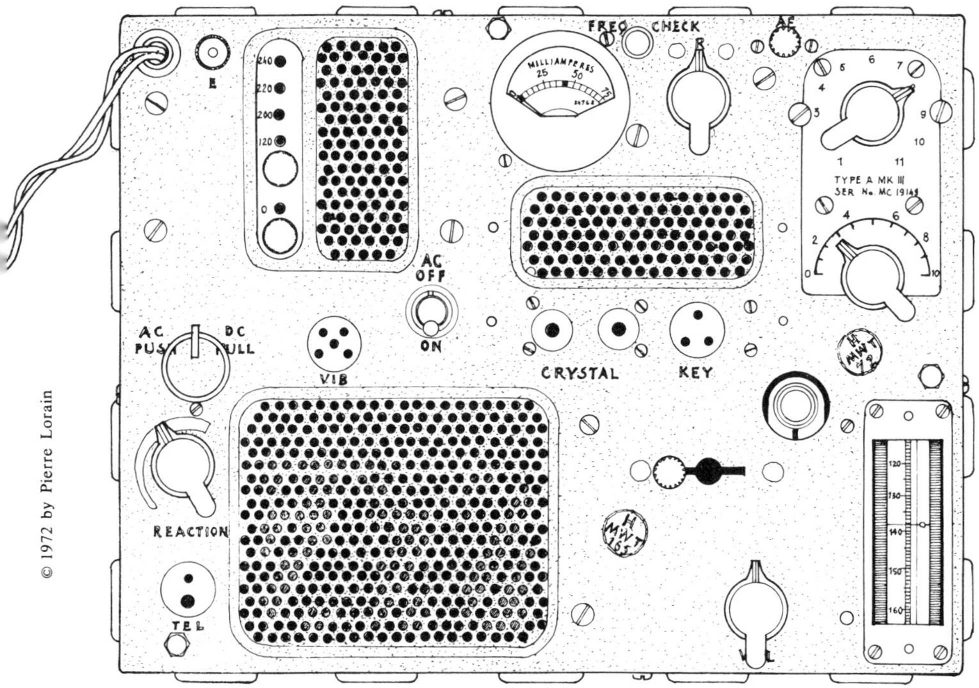


STANDARD FEATURES

*Coverage: 3.2–9 megacycles in 2 wave bands: "blue" band, 3.2–5.2 megacycles, "red" band, 5–9 megacycles*
*Power output: 5 watts average*
*Transmitter with 2 tubes, quartz pilot, and amplifier*
*Superheterodyne receiver with 4 tubes*
*Power supply of 100–250 volts, house current*
*In a single steel cabinet, black wrinkle finish, white lettering*
*Dimensions: 8½ × 7½ × 3¼ inches*
*Weight: About 5½ pounds*
*Micro key capped with black plastic*

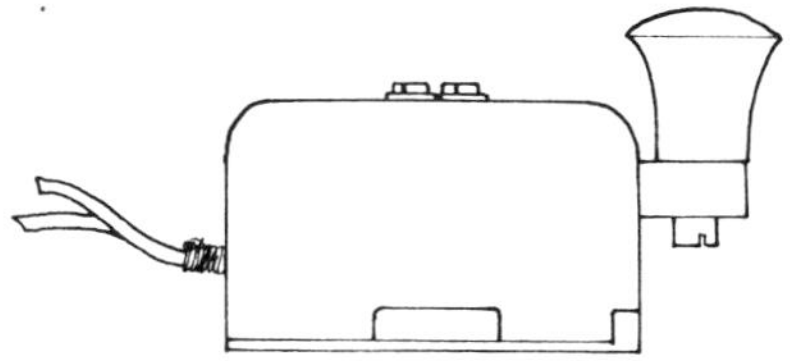

Half size

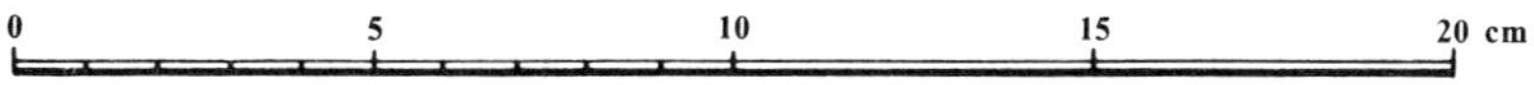


NOTE: *The complete set came with an accessories box and was housed in a very small suitcase, measuring 13 by 9 by 4 inches. The total weight was approximately 9 pounds.*

*On the set's front panel may be seen, from left to right and from top to bottom:*

*At the top: The double-twisted cord leading to the mains (house current) power supply, the ground terminal, the voltage selector, the milliammeter for transmitter tuning, the frequency control push button (marked* Freq-Check*), the transmission/reception switch, the aerial terminal, and the small plate engraved with the serial number, holding the 2 transmitter tuning knobs.*

*In the middle: The house current/battery (A.C./D.C.) switch, the vibrator socket, the main power switch, the quartz crystal holder, the key socket, and the neon frequency-control tube.*

*At the bottom: The superheterodyne control, the headset socket, the receiver wave-band switch, the volume control, and the receiver dial with 2-speed tuner. Two calibration curves show the frequency correspondences for the "blue" and "red" bands.*

# THE MINIATURE SETS

## 53 Mark I Receiver

### *Extra-flat Pocket Model*

NOTE: *Receiver was delivered in a canvas sack which also contained the power supply, earphones, 15 feet of aerial wire, 6 feet of ground wire, a lamp-socket outlet, and calibration curves. The complete set weighed about 2½ pounds.*

*Sketch is very slightly smaller than actual size.*

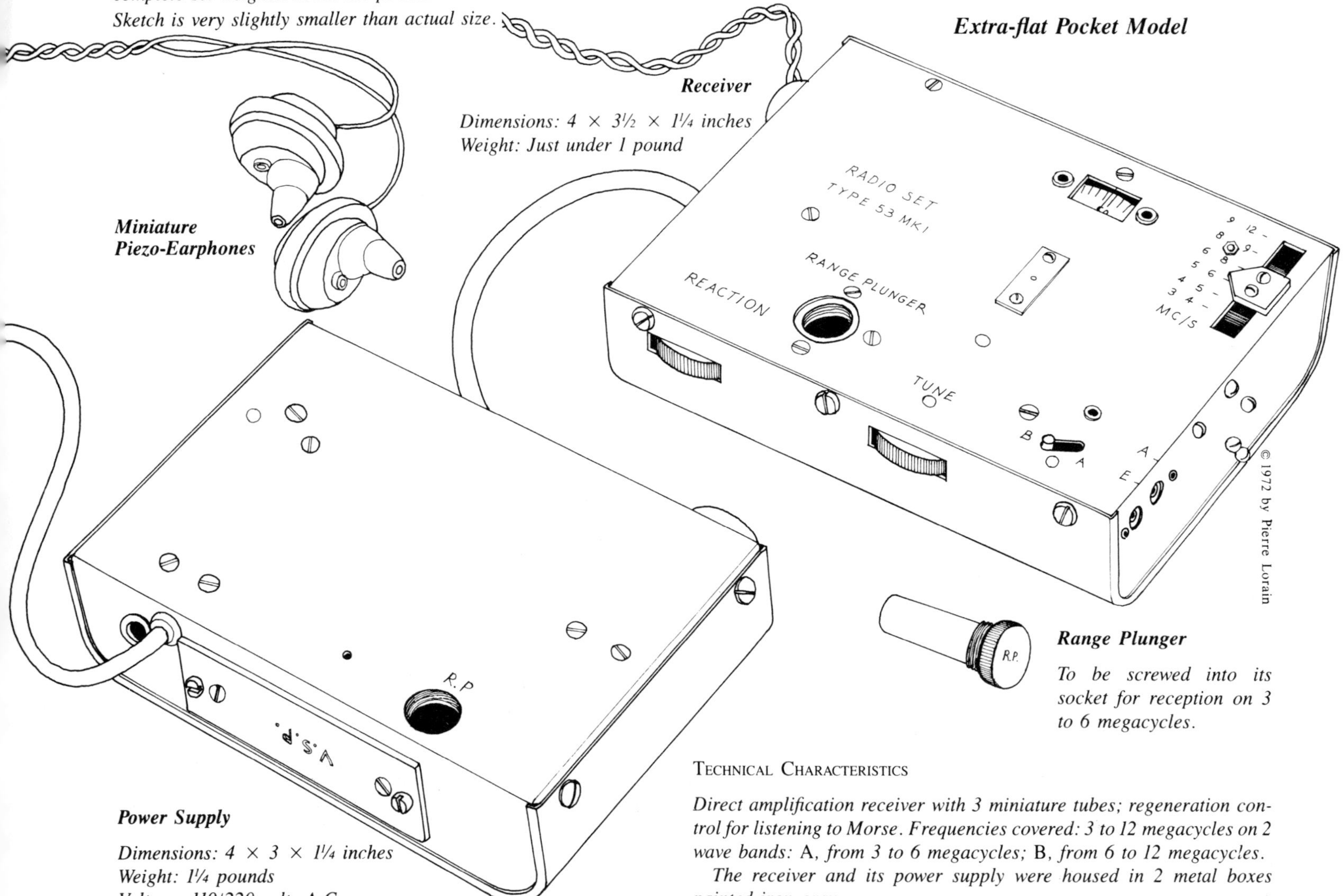


**Receiver**

*Dimensions: 4 × 3½ × 1¼ inches*
*Weight: Just under 1 pound*

***Miniature Piezo-Earphones***

**Range Plunger**

*To be screwed into its socket for reception on 3 to 6 megacycles.*

**Power Supply**

*Dimensions: 4 × 3 × 1¼ inches*
*Weight: 1¼ pounds*
*Voltage: 110/220 volts A.C.*

TECHNICAL CHARACTERISTICS

*Direct amplification receiver with 3 miniature tubes; regeneration control for listening to Morse. Frequencies covered: 3 to 12 megacycles on 2 wave bands:* A, *from 3 to 6 megacycles;* B, *from 6 to 12 megacycles.*

*The receiver and its power supply were housed in 2 metal boxes painted iron gray.*

# M.C.R. 1 Receiver ("Biscuit Receiver")

*Half size*

*Designed during the second half of the war, the "Biscuit" was a full-range, highly sensitive, superheterodyne receiver with 5 miniature tubes.*

*It was powered by a special battery, or with a separate power supply, by A.C./D.C. of 97 to 250 volts.*

*This receiver was delivered in a 2-pound "Huntley-Palmer" cookie tin which also contained 4 interchangeable coil units, 3 batteries or 2 batteries and a power supply pack, 30 feet of aerial wire, and 10 feet of ground wire.*

*A battery provided an average of 30 hours' reception time.*

*Four knobs on the same side controlled aerial adjustment, sensitivity, regeneration for telegraph reception, and the frequency dial.*

*Dimensions: 3½ × 2 × 8½ inches*

*Weight: About 2 pounds*

*Ten thousand of these sets were built, and approximately half were widely distributed in the clandestine networks.*



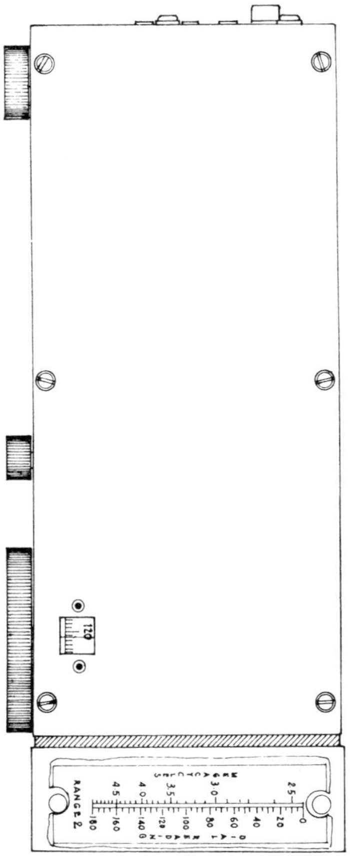


**Side View of the Receiver**

*Unit 2, covering 2.5 to 4.5 megacycles, is plugged in. The graduation shown on the dial corresponds to reception on a frequency of 3.6 megacycles. At the other end of the set are sockets for the headset, the aerial, the ground wire, and the power supply.*

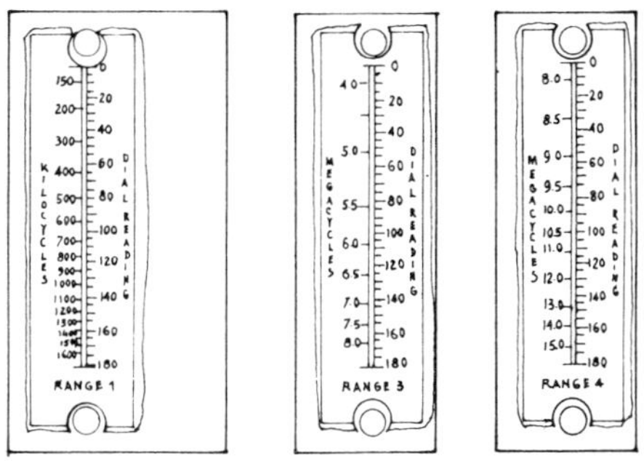


*These are interchangeable coil units to be plugged onto the pins at the end of the receiver.*

*A sticker on each unit shows the correspondence between the dial graduations and the frequencies received.*

*Unit 1 covers 150 kilocycles through 1.6 megacycles. This corresponds to the long and medium waves received on broadcast sets. Unit 3 covers 4 to 8 megacycles, and Unit 4, 8 to 15 megacycles.*

0 5 10 15 20 cm

## 51/1 Transmitter

***Three-Quarter Front View***

Standard Features

*Coverage: 3 to 10.5 megacycles on 2 wave bands: a) with range plunger in place: 3 to 5.8 megacycles; b) with range plunger removed: 5.8 to 10.5 megacycles*
*Power output: 3 to 4 watts*
*Transmitter with 3 miniature tubes, crystal-controlled oscillator, and current-rectifier tubes*
*Built-in 200 to 260 volt A.C. power supply pack*
*Steel box, black wrinkle finish*
*Dimensions: 5¾ × 4½ × 1½ inches*
*Weight: 1¼ pounds*
*Built-in micro key*

Preliminary Adjustments

*For a 200 to 220 volt A.C. supply, the 2 bridges were to be plugged into the* LO *sockets.*

*From 220 to 240 volts, one was to be plugged into the* LO *socket, the other in the* HI *socket.*

*From 240 to 260 volts, both bridges were to be plugged into* HI.

*The power supply, the ground, the aerial, the quartz, and the prong of the antenna coupler were plugged into their respective sockets. The key was depressed and the righthand knob was turned until the neon tube lit up. The antenna coupler pin was inserted into each successive socket until the neon tube lit up most brightly. The set was then ready to operate.*

*Operators who worked at the English home receiving stations still remember the excellent signals received from these tiny instruments, which sometimes maintained contact over a distance of 600 miles.*

***Range Plunger***

Note: *The transmitter was delivered with its "scindex" wire supply cord wrapped around the housing. Forty feet of aerial wire and 6 feet of ground wire were rolled up in the hollow between the 2 bakelite knobs on the front face. All this was in a canvas sack which also held 2 voltage selector bridges, a lamp-socket outlet, an adapter for regular-size quartz crystals, and perhaps any crystal or crystals needed.*

*An agent could thus continually change his transmission site carrying his entire set in his pocket. A 220-volt power supply sometimes presented difficulties when used in Paris, where several areas were still using 110 volts.*

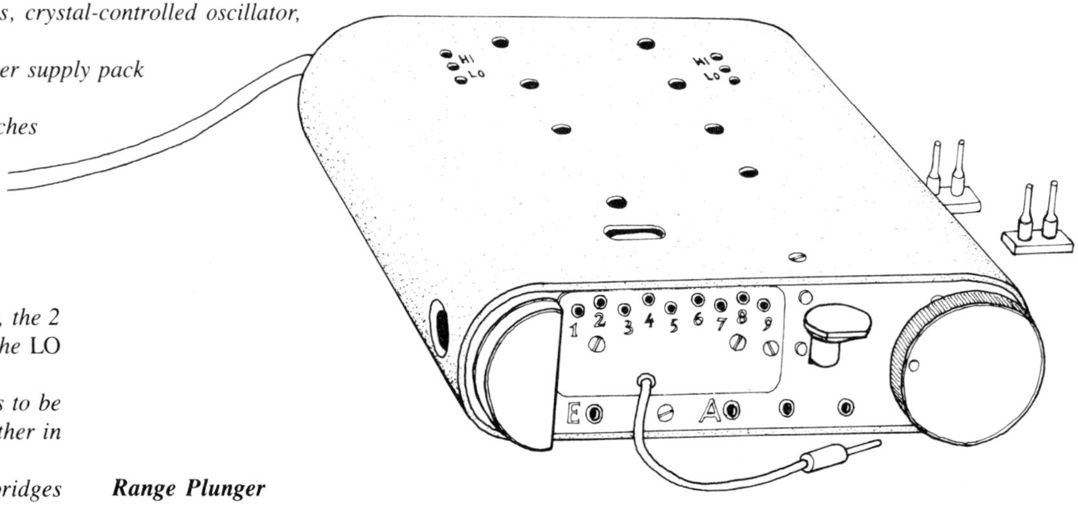

## *Extra-Flat Pocket Model*

*Half size*

*Top to bottom, the HI and LO sockets for voltage adjustment of a 200-to-260-volt power supply; ventilation holes, and an elongated opening at the bottom for observing neon tube brightness during transmitter tuning.*

*On the left is the grooved knob of the range plunger.*

***Top View***

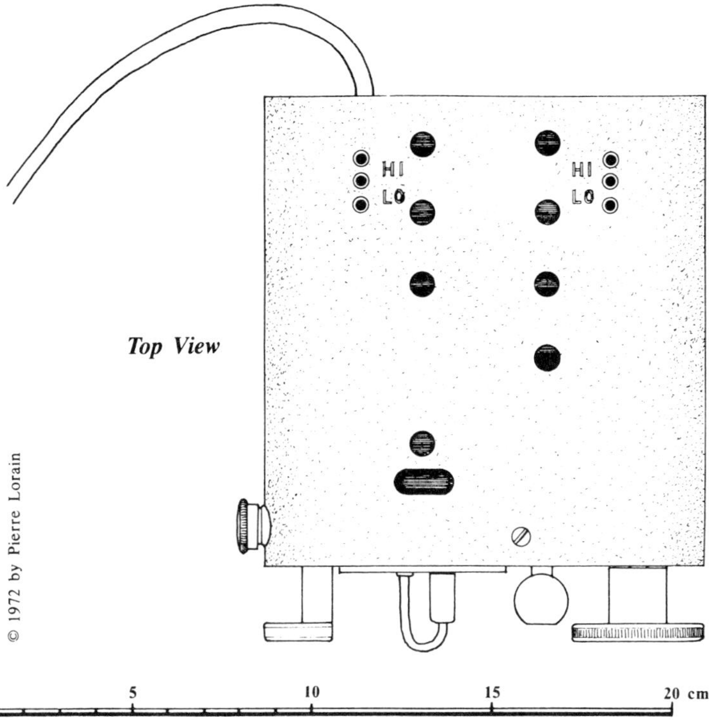




***Front View***

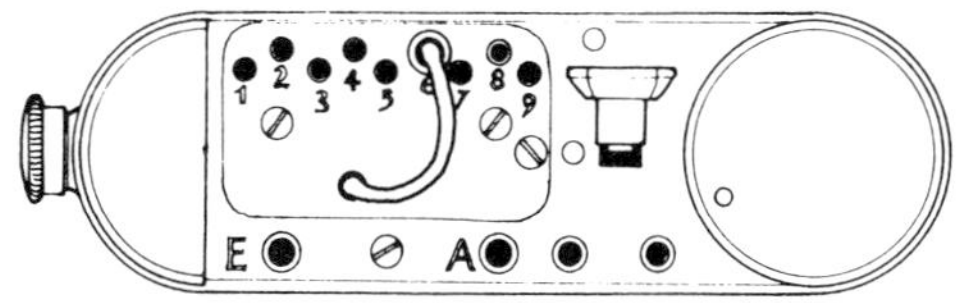


*Left to right, sockets for aerial circuit tuning,* E *socket for ground (earth) wire,* A *socket for aerial wire, the key, and the quartz crystal holder. On the right is the anode circuit adjustment knob.*

*American-made miniature quartz crystal*

***Detail***

*Actual size*

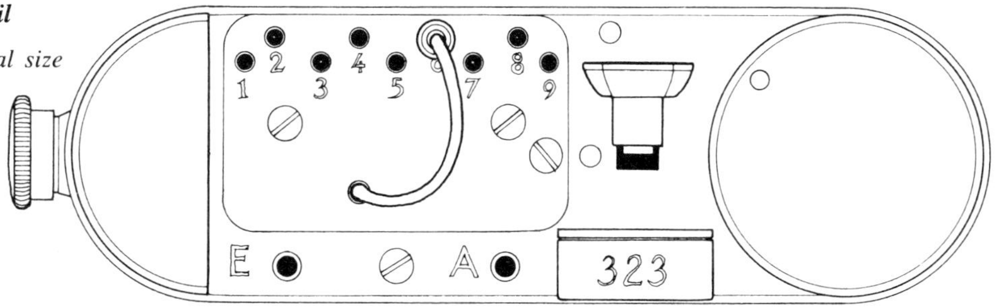


*The range plunger and quartz crystal are in place. After connecting with the aerial and, if necessary, the ground, the transmitter is ready to operate. If the quartz crystal frequency is greater than 5,800 kilocycles, the range plunger must be removed.*

## BATTERY CHARGERS

To deal with the growing frequency of power cuts in the months before the landing, the utilization of batteries quickly became standard practice. Recharging batteries was a critical problem. On request, the S.O.E. dropped seven types of battery rechargers:

- Thermocouple recharger. This consisted of 350 chromium-constantan cells in a fireproof brazier. With a good fire fueled by branches or charcoal, a 6-volt battery could be charged to deliver up to one ampere.
- House current recharger. The classic model.
- Wind recharger. This consisted of a small windmill mounted on a collapsible 10-foot-high pole. It would turn a generator capable of charging a battery to between 2 and 20 ampere capacity, depending on wind velocity.
- Crank-operated recharger. This could be set up on a table.
- Recharger that could be adapted to a bicycle. Average pedaling speed could charge a battery up to 5 amperes.
- Gasoline-powered recharger. The most common was the RB-8, composed of a small 2-stroke motor coupled to a generator. This could charge one 12-volt battery, or two 6-volt batteries, to 6 or 7 amperes.
- Steam recharger. This was the most amazing of all. It consisted of a boiler suspended over a brazier and attached by a flexible tube to a small 2-cylinder steam engine. The engine was coupled to a generator. The total weight did not exceed 70 pounds. With steam pressure at 30 pounds per square inch, an hourly consumption of 2 quarts of water, and 10 to 20 pounds of wood, a 6-volt battery could be charged to 4 amperes.

This recharger was used frequently in the Indo-Chinese and Burmese jungles in 1945 by Force 136. In this way, a group of saboteurs could remain isolated from all civilization for several months but still maintain a permanent radio link as long as they had some dry wood.

# Chapter IV

# Esoteric Links: Ciphers

*Cipher is one of the main components of any secret service. The survival of a network often depended on the impenetrability of its ciphers.*

*Although the subject is captivating, it is difficult to understand, and we will not bear a grudge against those who leaf through these pages. Most S.O.E. agents regarded their study of code as a dull school exercise!*

*We believe, however, that we should advise inquisitive minds to think of this chapter as intellectual recreation or a game of chess.*

*For these, we offer long hours of enjoyment.*

# GENERALITIES

### Principles of Classical Cryptography

In all armies, a coded telegram consists of a series of letters or numbers, generally grouped in fives to facilitate both counting these groups and the operator's task.

Although the systems used to transform a plaintext into a ciphertext are numerous and often complex, the principles which govern them are exceedingly simple and only 2 in number: transposition and substitution.

Let's look at transposition first. It is the simplest that comes to mind. Transposition is a rearrangement of the letters of the plaintext. In other words, it is an anagram.

For example, the word *radio* ciphered by transposition can appear as AROID, ROADI, IDARO or any other combination which results from rearranging the 5 letters. It is obvious that the security afforded by a transposition system is directly proportional to the length of the text. A text of less than 150 characters would only offer illusory security.

The second principle is substitution. Each letter of the plain alphabet is replaced by another letter, number or design. For practical reasons, this is done by using a so-called cipher alphabet, which lists these replacements.

For example:

| Plain alphabet | | | | Cipher alphabet |
|---|---|---|---|---|
| | *a* | = | I | |
| | *b* | = | D | |
| | *c* | = | F | |
| | *d* | = | R | |
| | *e* | = | W | |

Therefore, the word *cab* coded with the substitute alphabet will read FID.

If only one cipher alphabet is used, a plaintext character will always wear the same disguise. This monoalphabetic procedure does not, of course, offer security.

The more usual substitution systems are all polyalphabetic. This means that instead of each plaintext letter being represented by only one cipher character, it can be replaced by any one of the letters in an entire alphabet. Therefore, each letter can assume 26 different disguises. Which one employed is specified by a key (a succession of characters or numbers agreed upon in advance).

For example:

| Key | | A | B | C | D | E | F | G | H | I | J | . | |
|---|---|---|---|---|---|---|---|---|---|---|---|---|---|
| Plain alphabet | *a* = | D | K | U | F | D | R | M | E | W | S | . | Cipher alphabets |
| | *b* = | S | D | E | W | R | O | T | . | . | . | . | |
| | *c* = | U | A | R | F | T | . | . | . | . | . | . | |

With this arrangement (a form of the Vigenère tableau), it is easy to group the cipher alphabets that correspond to each character of the plain alphabet. The key at the top determines the cipher letter to be used. In other words, each cipher letter has as coordinates, on the one hand, the corresponding plain letter and, on the other, the corresponding key letter.

*Example*. Assume that the word CAB is the key selected and, to simplify this demonstration, that plaintext is composed only of the letter *a*. The encipherment setup would appear as follows:

| | | | | | | | | | | |
|---|---|---|---|---|---|---|---|---|---|---|
| Key | C | A | B | C | A | B | C | A | B | C |
| Plaintext | *a* | *a* | *a* | *a* | *a* | *a* | *a* | *a* | *a* | *a* |
| Ciphertext | U | D | K | U | D | K | U | D | K | U |

Therefore, a single character will assume 3 disguises (U, D, and K).

If the key is comprised of 10 different letters, the plain letter will take on 10 different disguises. At the most, all the letters of the cipher alphabet will successively replace the plain character when a key with all 26 letters is used. This would guarantee maximum secrecy.

Therefore, the security of this system is directly proportional to the number of characters which comprise the key and the incoherence of the letters that compose it.

Up to this point we have only looked at letter-for-letter, or monographic, substitution. Two-letter-for-two-letter, or bigraphic, substitution can also be considered. Bigraphic substitution lists like the following can be laid out:

| Plain bigrams | | Cipher bigrams |
|---|---|---|
| *es* | = | TR |
| *et* | = | DK |
| *eu* | = | LF |

Such lists should be frequently revised. For with a single list, the same cipher bigram always represents the same plain bigram. However, this system can be made safe enough if there are scores of numbered lists which can be employed in succession, according to an agreed-upon key.

Carrying the principle of replacing more than one letter at a time to its ultimate, one can envisage replacing a word, a part of a sentence, or even an entire sentence with a group of characters or numbers. This offers the advantage of reducing the length of the ciphertext. Thus we arrive at codes, which consist of lists of the plaintext words and phrases and their code equivalents. Codes may therefore be viewed as nothing more than a sort of polygraphic substitution.

Thus, for the following plaintext, "*Extremely urgent drop/nine/type H5 containers/on ground area K,*" 4 groups of letters or numbers may be substituted. The coded message might read, "KBSTR TDEFG BRIWD AKBDS" or "65421 43246 25352 10456."

## Superencipherment, or Enciphered Code

A code in which each word or phrase is always represented by the same code group offers only a fairly low level of security since certain expressions are often repeated during a series of messages. One solution is to assign several different code groups to the same expression.

Since a code can be captured, it would be wise to at least camouflage the groups before transmission. This concealment, called "superencipherment," can be accomplished by applying the transposition or substitution procedures we have just considered to the code groups;

or, more simply, if it is a code using numerical groups, by adding (without carrying) a superencipherment key, called an "additive key," to these groups.

Therefore, using the previous message as an example, with the superencipherment key 4932154, we have the following:

| | | | | |
|---|---|---|---|---|
| Code groups: | 65421 | 43246 | 25352 | 10456 |
| Additive key: | 49321 | 54493 | 21544 | 93215 |
| Superenciphered groups: | 04742 | 97639 | 46896 | 03661 |

To decipher with the same key, one subtracts without borrowing.

Code books are small vocabularies or dictionaries that contain the most common words or phrases in alphabetical order, along with enciphered equivalents (in groups of 3, 4, or 5 numbers or letters, depending on the size of the vocabulary).

Code books are divided into 2 parts, much like a bilingual dictionary. One part is used for encoding, the other for decoding.

Easy and quick to use, these books cut down on coding errors and do not require the encoder to give sustained attention to the task. They were widely used by all echelons during World War II.

## Automatic Cryptography: Cipher Machines

Each of the systems we just reviewed can, by themselves, serve as a valid base for enciphering messages. They can be complicated at will, for example, by transposing a plaintext before substituting or by superenciphering a code. However, it should be remembered that in the upper echelons of the military, the communications volume is quite substantial. Therefore, it is important to simplify as much as possible, or even automate, the heavy and unrewarding task of the cipher clerks.

Technological advances offered a solution. The first commercial cipher machines appeared on the market around 1920. Primitive and barely safe at first, around 1935 they reached a high level of efficacy.

Most were German- or Swiss-built and usually automated a polyalphabetic substitution system with an extremely long key (several hundred thousand incoherent series of letters). The smallest were manual and did not print out the message; the largest were electric and had a typewriter keyboard. All the cipherer had to do was type out the plaintext, and the machine printed the ciphertext in 5-letter groups.

If 2 of these machines were set to the same key—in other words, if both had the same basic key—they generated 2 identical keys that flowed automatically from the basic key. Typing out the plaintext on one machine printed out the ciphertext; typing out the ciphertext on the second machine printed out the plaintext. The cipher clerk's work was little more than typewriting.

Because the machine produced its key during the typing, setting the key would be the work of only one person; protection against a possible indiscretion was thus very great.

By 1936, most of the great powers possessed these machines. The smaller of these machines were not much larger than a portable typewriter.

Finally, during World War II, the transmission stations at the highest level had high-speed teletypewriting cipher machines. Some of these consisted of a standard teletypewriter hooked up to an automatic ciphering box. A text typed out in plaintext at the point of origin was automatically enciphered for transmission, then automatically deciphered during reception, the receiving machine typing out the plaintext as the message arrived.

## The Principles of Clandestine Cryptography

From among the procedures just described, let us pick out those which were usable in clandestine operations.

Teletypewriter cipher machines had to be eliminated for obvious technical reasons. These instruments, which were usually housed in heavy trailers, required a heavy power supply to operate and were necessarily linked to powerful transmitters. Cumbersome in the field, they were reserved for normal transmissions in the highest echelons.

We could next eliminate all other cipher machines.

Aside from their great ease of operation, they offered no cryptographic advantages over the corresponding pencil-and-paper method. Therefore, the possession of such a machine by an operator would be nothing more than a dangerous hindrance.

One was equally tempted to eliminate code books. The thick, dictionary type of code book would have been impossible to conceal. Such code books were meant to be kept in embassy vaults or warship safes. Nevertheless, they had the great advantage of considerably reducing the length of the ciphertext. Accordingly, transmission was shorter. For this reason, for clandestine communications, they could take the form of short glossaries: a few microfilmed pages, for example, or printing the commonest expressions of a particular type of operation on a piece of cloth the size of a handkerchief.

Polyalphabetic substitution would also be retained since only an alphabet grid and a memorized key were needed for this.

Transposition systems would be included since they required nothing more than a memorized key, paper, and pencil. They answered the necessity of supplying an agent with no more compromising articles than necessary.

It should be noted that although a text made up of 5-letter or 5-digit groups is perfect for radio communications, it becomes extremely compromising if transmitted in written form. In case of capture, no explanation was possible. Therefore, 2 specifically clandestine cryptographic systems are included here for written communications (postcards, letters, and packages sent by mail). These consist of invisible writing (sympathetic inks) and the concealment of the secret message in a letter with an innocent appearance.

#### The Notion of Secrecy

There are degrees of secrecy. These are expressed by the time during which a particular secret is to be kept.

It would be best if this duration could be infinite, but this can only be achieved under very particular circumstances.

Therefore, it is fitting that, in standard cryptography, secrets be put into categories. Diplomatic correspondence must be kept secret for months; if possible, for years.

During wartime, strategic military communications between large units should be able to resist enemy solution for several weeks, and if possible, several months.

Finally, tactical communications between smaller commands should keep their secrets for several days.

To each of these categories of communications correspond several cryptographic systems. The first category requires the most highly developed systems employed by the high-security machines. The second category accepts less complex systems and machines, and the third, simple systems which are quick and easy to use.

Secret services require, for themselves, practically absolute security, so that a secret might possibly be kept forever. Since the use of big codes or of machines is prohibited under certain circumstances, the problem seems difficult to resolve. We will see how the S.O.E. did it throughout its existence.

## THE CRYPTOGRAPHIC SYSTEMS OF THE S.O.E.

### Two-Way Radio Communications Between the Home Station and the Network

#### The Playfair Cipher of the First Months (1941)

The first system was nothing more than a revival of the British army's field cipher from World War I. It was called the Playfair (after Sir Lyon Playfair, an English scientist and member of Parliament, who had recommended using the system in 1854 during the Crimean War, though it was actually one of the many inventions of the ingenious physicist, Charles Wheatstone).

It offered the advantage of not leaving any compromising item in the hands of the agent, who left on assignment with only a few lines of poetry in his head.

It was simple bigraphic substitution. The ingenuity of the invention consisted of replacing the list of 676 possible bigrams ($26^2$) by a 25-letter square which was easily drawn up with the help of a mnemonic method. A typical example is:

The memorized verse: "Mary had a little lamb; its fleece was white as snow."

Square formation: The verse is written horizontally into a 5 × 5 square, with letters repeating previous ones omitted. The square is then completed by listing, in alphabetical order, the letters that did not appear in the verse. I and J are put in a single cell:

| | | | | |
|---|---|---|---|---|
| M | A | R | Y | H |
| D | L | IJ | T | E |
| B | S | F | C | W |
| N | O | G | K | P |
| Q | U | V | X | Z |

Encipherment: If the word *England* is to be enciphered, it is first broken down into bigrams: *en gl an dx*

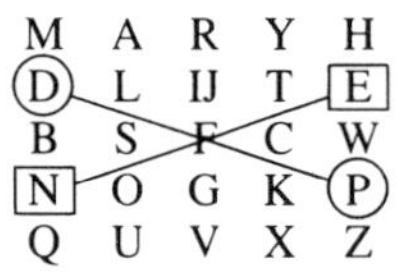


If the letters of the plain bigram form the diagonal of a rectangle in the square, the other diagonal of the same rectangle would yield the cipher bigram.

Therefore, *en* is ciphered DP. Likewise, *gl* is ciphered OI or OJ. Finally, the rectangle created by diagonal *an* has MO for its other diagonal. The cryptogram will read: DP OJ MO TQ.

When the 2 letters of the plain bigram do not form a diagonal but stand on the same line, each letter is replaced by the letter to its right. Therefore, the plain bigram *ti* would be enciphered as ET. If the 2 letters of the bigram stand in a column, each letter is replaced by the letter beneath it. Thus the plain bigram *lu* would be enciphered as SA.

Decipherment works in reverse.

The advantages of the Playfair cipher are that no compromising documents are required and that it is easy and very quick to use.

Its disadvantages are that with a given square, the same cipher bigram will always represent the same plain bigram. Therefore, the frequently used French bigram *es* will always be enciphered as LW on the square just laid out.

Consequently, a long message or series of messages enciphered with the same square will bring about repetitions which will facilitate analysis, and finally, solution by the enemy.

Such a cipher system can only be good if the square is frequently changed.

Starting in 1942, the S.O.E. restricted the use of Playfair to internal network communications and prohibited its use for any message transmitted by radio. The Playfair was then discarded in favor of transposition systems.

### The Double Transposition (1941 to 1942)

Although the principle of this system is simple, its application is long and painstaking.

Message to be enciphered: *Urgent. Send munitions* with the first numerical key 251463.

Simple transposition: The plaintext is written out horizontally under the first numerical key:

| 2 | 5 | 1 | 4 | 6 | 3 |
|---|---|---|---|---|---|
| *u* | *r* | *g* | *e* | *n* | *t* |
| *s* | *e* | *n* | *d* | *m* | *u* |
| *n* | *i* | *t* | *i* | *o* | *n* |
| *s* | | | | | |

The letters are then transcribed vertically by columns in their numerical order and recorded horizontally: GNT USNS TUN EDI REI NMO.

To mix the letters even more, the encipherer repeats the operation with a second numerical key different from the first, for example, 7312645.

Double transposition: The first cryptogram is written in horizontally under the new key:

| 7 | 3 | 1 | 2 | 6 | 4 | 5 |
|---|---|---|---|---|---|---|
| G | N | T | U | S | N | S |
| T | U | N | E | D | I | R |
| E | I | N | M | O | | |

The letters are transcribed vertically in the numerical order of the columns and recorded horizontally to give the final cryptogram: TNN UEM NUI NI SR SDO GTE.

The letters are then grouped in fives. The cryptogram is then ready for radio transmission.

Deciphering is carried out in inverse order. The cryptogram is entered vertically by columns under the second numerical key. Then it is extracted horizontally. The result is inscribed vertically under the first numerical key. The plain telegram is read horizontally.

In order to make this system safer, it was taught in secret S.O.E. classes (as late as the beginning of 1942) that:

- A telegram should have a minimum of 100 characters and a maximum of 400 to 500.
- The 2 numerical keys should be of different lengths, for example, 10 and 12 digits, 8 and 11, etc.
- Seven to 10 nulls, or dummy characters, should be added at the beginning as well as enough at the end to form a message whose number of characters is a multiple of 5, while avoiding filling out the entire transposition table.

The keys changed for each message. During the initial months of hostilities, these were determined by using a book, such as a novel; both the agent and the home station had a copy of the same edition. Two groups placed at the head of the message indicated the page number, the line, and the number of letters to be used for the two tables.

Since in the cryptogram these 2 groups would have to be made up of letters and not numbers, it was necessary to use a table for converting numbers into letters. But first, as a security measure, the operator superenciphered these groups by adding to them, without carrying, his secret identification number.

It is difficult to imagine that people could have seriously thought of using such a complicated system.

It should be pointed out that the unhappy operator had to retain a book and memorize a number-to-letter conversion table and the secret number to superencipher the first 2 groups, all at the same time. It was a real nightmare. A complete illustration of this system is given in the appendix.

The principle of using a book to determine the keys was soon recognized as impractical by the British networks and replaced by a memorized verse. This example comes from the cryptography course taught in the secret school at Beaulieu, near Southampton, at the start of 1942.

The 2 transposition keys were determined from a poem, one of those known to have been memorized by every French schoolchild. But for this example in English, suppose the lines to be (instead of poetry) the version used during World War II of the pledge of allegiance: "I pledge allegiance to the flag of the United States of America and to the republic for which it stands, one nation, indivisible, with liberty and justice for all."

1. Each word is represented by 1 letter of the alphabet:

| I | PLEDGE | ALLEGIANCE | TO | THE | FLAG | OF | THE | UNITED | STATES | OF | AMERICA |
|---|---|---|---|---|---|---|---|---|---|---|---|
| A | B | C | D | E | F | G | H | I | J | K | L |
| AND | TO | THE | REPUBLIC | FOR | WHICH | IT | STANDS | ONE | NATION | INDIVISIBLE | |
| M | N | O | P | Q | R | S | T | U | V | W | |
| WITH | LIBERTY | AND | | | | | | | | | |
| X | Y | Z | | | | | | | | | |

2. The operator chose 6 consecutive letters at random: for example, JKLMNO.

It was arbitrarily decided that the odd letters would furnish the first key, and the even letters, the second.

Therefore, the first key became JLN and read: "STATES AMERICA TO."

The second key, formed by KMO, reads "OF AND THE."

3. For each key, a number was assigned to each letter according to its alphabetical order:

| | | | | | | | | | | | | | | | |
|---|---|---|---|---|---|---|---|---|---|---|---|---|---|---|---|
| *First key:* | S | T | A | T | E | S | A | M | E | R | I | C | A | T | O |
| | 11 | 13 | 1 | 14 | 5 | 12 | 2 | 8 | 6 | 10 | 7 | 4 | 3 | 15 | 9 |

| | | | | | | | | |
|---|---|---|---|---|---|---|---|---|
| *Second key:* | O | F | A | N | D | T | H | E |
| | 7 | 4 | 1 | 6 | 2 | 8 | 5 | 3 |

4. After enciphering his message, the operator had to indicate, in the second group (the first being the identification group), the keys he selected.

The 2 keys are perfectly defined by the letters J, L, and N. Two nulls were inserted among them to form a 5-letter group: J D L X N.

To conceal this group, each of its letters was replaced by the second letter following it in the alphabet:

| | | | | |
|---|---|---|---|---|
| J | D | L | X | N |
| K | E | M | Y | O |
| L | F | N | Z | P |

The group for the message became LFNZP.

5. The home station reversed the procedure to decipher the message.

This system was a simplification of the original one. The book and the secret number (which agents could not always remember) were done away with. Nevertheless, this system was still too complicated and was replaced, after several months, by the "A–Z Code." This was a substantial improvement.

### The A–Z System

This was nothing more than a double transposition system, but with the keys prepared in advance and printed on a silk handkerchief given to the agent at the moment of his departure.

There were 2 series of keys. One fell under the heading "Your keys to us." The agent used this series to encipher his messages for the home station. The other was called "Our keys to you" and was used to decipher messages sent from the home station.

Each pair of numerical keys was preceded by a 5-letter indicator group. Before transmission, the encipherer wrote it at the beginning and end of the message. It told the home station which lines to choose for deciphering.

The advantages over earlier ciphers are obvious. Since the keys were ready to use, the agent no longer had to remember a secret number or conversion table. Consequently, ciphering errors were substantially reduced. The bulky, dangerous book was abandoned. The keys, consisting of long number series, were scrambled by machines in an incoherent manner, increasing the system's security.

Finally, an agent subjected to torture was not able to reveal anything since he had not memorized anything.

On the other hand, it was dangerous to hold onto the handkerchief. After enciphering a message, it was imperatively specified that the lines of the key just used be cut off the handkerchief and destroyed immediately. This was to avoid, in the event of the agent's capture, the enemy's reading the messages which might have been recorded by his radio monitoring networks. Unfortunately, silk does not burn easily.

In the appendix, there is a reproduction of microfilm directions for using the A–Z Code, drawn up in French for B.C.R.A. agents.

### Criticisms of the A–Z System

Even with ready-to-use keys, double transposition engenders ciphering errors. If a column is transcribed out of order, an entire portion of the text will be undecipherable or require long hours of work to rediscover the plaintext.

Agents who had to use this system in the early years detested it. One of the most important and earliest British agents still remembers "the long hours spent helping

| | |
|---|---|
| BKDFH | 17 3 2 9 8 7 14 11 6 10 4 13 16 12 15 1 5<br>1 4 16 8 24 6 15 25 9 10 23 26 5 17 28 11 27 2 18 22 29 12 19 7 14 20 21 13 3 |
| KDEUP | 3 15 7 11 21 1 14 2 9 13 6 20 12 8 17 4 18 10 19 16 5<br>8 4 12 1 6 11 5 13 15 2 7 14 9 10 3 |

the radio operator decipher, or try to decipher, the messages received!"

In addition, we have seen that, to provide adequate security, all telegrams enciphered by transposition should have had no fewer than 100 letters. Greater security was attained with 150 letters. This meant that the radio operator had to send out at least 30 groups, even if the information he had to transmit was contained in only 2 or 3 groups.

There was a more serious problem than this, however. Even though deciphering double transposition is difficult, it is mathematically possible by classic cryptanalytical methods, given certain conditions.

In the summer of 1942 the Allies began to use the first IBM tabulators on a large scale for cryptanalysis. The results were spectacular: many enemy ciphers were broken within a few weeks. The famous double transposition system could not resist any more than other systems, and probably even much less.

Since it was logical to assume that the Germans were using similar equipment at the same time (which they were), this system, inferior and just barely safe in principle, was condemned at the end of 1942. However, for security and administrative reasons, it was not possible to replace double transposition overnight with the simple, inviolable, universal system which was forthcoming. (Called the "one-time pad," the S.O.E. F Section was not to use this system regularly until September 1943.)

Unfortunately, the change could only take place slowly and in stages. Also, to improve security, several other systems were introduced along the way. We will regard these as transitional systems.

### Transitional Systems (1942–1943)

The systems of transition were used in conjunction with double transposition (slowly dying out in the form of the A–Z cipher) between November 1942 and August 1943.

Double transposition was condemned in the technical sense. Its condemnation included *ipso facto* all other transposition systems. It was therefore necessary to return to the substitution principle.

Since the Playfair did not offer sufficient security for use by radio, the English adopted the logical temporary solution: the Delastelle system. It was named after its inventor, Félix-Marie Delastelle, who described the principle for the first time in his book, *Cryptographie Nouvelle* ("New Cryptography"), published in 1893.

This is one of the most difficult hand systems to solve. Another advantage is that no message is needed.

## The Delastelle System

Using the same alphabet square that we used to explain the Playfair, let us place the digits *1* through *5* along its sides:

| | 1 | 2 | 3 | 4 | 5 |
|---|---|---|---|---|---|
| 1 | M | A | R | Y | H |
| 2 | D | L | IJ | T | E |
| 3 | B | S | F | C | W |
| 4 | N | O | G | K | P |
| 5 | Q | U | V | X | Z |

Sentence to be enciphered: *Send exploding coal.*

The plain text was normally broken up into groups of 5 letters. The coordinates of each letter of the plaintext ($S$ = 32, $E$ = 25, etc.) were written vertically:

| | | | | | | | | | |
|---|---|---|---|---|---|---|---|---|---|
| *s* | *e* | *n* | *d* | *e* | *x* | *p* | *l* | *o* | *d* |
| 3 | 2 | 4 | 2 | 2 | 5 | 4 | 2 | 4 | 2 |
| 2 | 5 | 1 | 1 | 5 | 4 | 5 | 2 | 2 | 1 |
| S | O | L | Q | H | X | T | T | U | D |

The coordinates that will create the cipher letters were read out, horizontally—32, 42, 22, 51, 15. The corresponding cipher letter was found in the grid: 32 = S, 42 = O, 22 = L, etc.

Therefore, the cryptogram reads: SOLQH XTTUD TGGRO AZUTK.

Decipherment was carried out by reversing the operation. The cipher letter coordinates were written horizontally and the vertical coordinates determined the corresponding letters which were found in the square.

| | | | | | | | | | | |
|---|---|---|---|---|---|---|---|---|---|---|
| S | O | L | Q | H | X | T | T | U | D | . . . |
| 3 | 2 | 4 | 2 | 2 | 5 | 4 | 2 | 4 | 2 | . . . |
| 2 | 5 | 1 | 1 | 5 | 4 | 5 | 2 | 2 | 1 | . . . |
| *s* | *e* | *n* | *d* | *e* | *x* | *p* | *l* | *o* | *d* | . . . |

The Delastelle system was one of the best. However, analysis was still possible since this system is not mathematically unsolvable.

For a cryptogram to be unsolvable, it must be able to assume every possible meaning so that an infinity of solutions is possible. In this way, it will escape analysis. To realize this, an incoherent variable must be introduced into the system, that is, an infinitely long key made up of an incoherent series, produced at random, of numbers or letters.

From the Delastelle system logically follows the idea of using such an incoherent numerical key, for example, with an additive system without carrying, whose principle we examined in superenciphering numerical codes. This leads to the principle of the second system, called the "numerical system."

| | | | | | | | | | |
|---|---|---|---|---|---|---|---|---|---|
| *i* | *n* | *g* | *c* | *o* | *a* | *l* | *x* | *x* | *x* |
| 2 | 4 | 4 | 3 | 4 | 1 | 2 | 5 | 5 | 5 |
| 3 | 1 | 3 | 4 | 2 | 2 | 2 | 4 | 4 | 4 |
| T | G | G | R | O | A | Z | U | T | K |

### The Numerical System

The Delastelle system square was employed, but each side was numbered twice, using the digits 0 through 9 instead of just 1 through 5 (in other words, the coordinates of Y, for example, were 14, 19, 64, or 69).

| | 1/6 | 2/7 | 3/8 | 4/9 | 5/0 |
|---|---|---|---|---|---|
| 1/6 | M | A | R | Y | H |
| 2/7 | D | L | IJ | T | E |
| 3/8 | B | S | F | C | W |
| 4/9 | N | O | G | K | P |
| 5/0 | Q | U | V | X | Z |

Assume that the key is a series of numbers picked at random: for example, 2156 4207 3109 0036 4977 3694 0735 1618 . . . Sentence to be enciphered: *Send exploding coal.*

The text was usually broken down into 5-letter groups. One of the 4 possible coordinates was entered under each plain letter. Under each of these coordinates, the numbers of the keys were written in order.

Adding (without carrying) these (plain letter) coordinates to the key numbers gave the cipher letter coordinates. The corresponding cipher letter was found in the square: 53 = V, 76 = D, etc.

Therefore, the cryptogram read: VDFJQ VPVBG, etc.

| | | | | | | | | | | | |
|---|---|---|---|---|---|---|---|---|---|---|---|
| *Plain* | *s* | *e* | *n* | *d* | *e* | *x* | *p* | *l* | *o* | *d* | . . . |
| *Plain coordinates* | 32 | 20 | 91 | 21 | 20 | 04 | 40 | 22 | 42 | 21 | . . . |
| *Key* | 21 | 56 | 42 | 07 | 31 | 09 | 00 | 36 | 49 | 77 | . . . |
| *Sum: cipher coordinates* | 53 | 76 | 33 | 28 | 51 | 03 | 40 | 58 | 81 | 98 | . . . |
| *Enciphered text* | V | D | F | J | Q | V | P | V | B | G | . . . |

Decipherment was easily accomplished by the inverse operation. The key numbers were subtracted from whichever set of ciphertext coordinates the decipherer happened to choose. (The plain coordinates of sender and receiver might differ, but the plaintext would be the same for both.)

| | | | | | | | | | | | |
|---|---|---|---|---|---|---|---|---|---|---|---|
| *Ciphertext* | V | D | F | J | Q | V | P | V | B | G | ... |
| *Cipher coordinates* | 58 | 76 | 88 | 78 | 56 | 58 | 95 | 58 | 86 | 98 | ... |
| *Key* | 21 | 56 | 42 | 07 | 31 | 09 | 00 | 36 | 49 | 77 | ... |
| *Difference: plain coordinates* | 37 | 20 | 46 | 71 | 25 | 59 | 95 | 22 | 47 | 21 | ... |
| *Plaintext* | *s* | *e* | *n* | *d* | *e* | *x* | *p* | *l* | *o* | *d* | ... |

This system is mathematically unbreakable if the key is never reused since the plaintext meaning of the ciphertext group depends on the key: if the key is changed, the meaning also changes. For example, the ciphertext in the preceding cryptogram can signify *Paris* instead of *sende* if the key is changed while the same square is used:

| | | | | | |
|---|---|---|---|---|---|
| *Ciphertext* | V | D | F | J | Q |
| *Cipher letter coordinates* | 58 | 76 | 88 | 78 | 56 |
| *New key* | 13 | 14 | 75 | 50 | 29 |
| *Difference: plain coordinates* | 45 | 62 | 13 | 28 | 37 |
| *Plaintext* | *p* | *a* | *r* | *i* | *s* |

Analysis is impossible since the cryptogram has all possible messages of the cryptogram's length as solutions and an infinite number of keys, all of the latter—from the point of view of the cryptanalyst—equally probable.

Accordingly, it was necessary to give the agent a silk handkerchief on which the key groups were printed, since—being random—they could be neither generated nor memorized. The agent had to have twice as many key numbers as letters to be enciphered. After using the key groups, he cut them off and burned them, as with the A–Z cipher. The groups could be used only once.

This system was a considerable improvement over previous ones. For the first time, a method was available that was absolutely inviolable. The security of radio transmissions became total.

Nevertheless, the application of this system was somewhat time-consuming from the point of view of both enciphering and deciphering. It required a certain amount of concentration, and although the addition and subtraction were extremely easy, they were the source of many errors. Numbers constituted an intermediate step better left out, if possible. The ideal situation consisted of going directly from the plaintext to the ciphertext using a Vigenère tableau and an incoherent key.

Therefore, all an agent needed was keys printed on

nitrated paper (since silk does not burn easily) and a Vigenère tableau printed on a silk handkerchief.

This is how we came, step by step, to the universal one-time pad system foreseen by the end of 1942. We will now take a look at it.

### The Ultimate Weapon: The One-Time Pad (or Key Used Only Once)

Starting in September 1943, until the end of the war, the entire F Section of the S.O.E. employed this system. This cipher represented the pinnacle of development in the field of cryptography. It is an extremely quick and easy system, does not require constant attention, and does not generate many errors. Finally, of prime importance is the fact that this cipher is mathematically unbreakable. No solution is possible, even with the fastest computers. Today, it is used for communications at the highest level where absolute security is required.

Like all great successes, the system is the embodiment of simplicity. The agent must have a table printed on silk and about 10 microfilm pages of keys. This slight risk is largely compensated by the enormous increase in security for the entire network. Furthermore, if the agent is captured, he will be unable to reveal anything since he did not memorize anything.

The cipher is a polyalphabetic substitution. Depending on the key, each plain letter can be replaced by any one of the 26 letters of the alphabet.

All the cunning of this system lies in the composition of this key. We saw that security was directly proportional to the length and incoherence of the key. The first polyalphabetic substitution systems employed short keys that were soon replaced by long and coherent keys (such as lines from a book). Each key was used for a short period of time.

In the case of the one-time pad, the key is made up of an incoherent and practically infinite series of letters. The length of the key corresponds to the length of the text to be enciphered. The key was burned as soon as ciphering was completed. Only the home station had a duplicate.

### The Principle of the One-Time Pad

- Type of tableau: key letters at the top. In the column under each key letter, the capital letters represent the plain alphabet, and the small letters, the cipher alphabet. (This is contrary to the typographic convention of this book, but is used here for historical accuracy.)
- One-time keys: for example, the following incoherent series: GDCBB WKNOC TRBAC KDZVU BRAWD . . .
- The text to be ciphered: SEND US A LYSANDER, which amounts to three 5-letter groups. The first 3 key groups are entered above the plaintext. The ciphertext will be found in the table opposite the plain letter in the column under the corresponding key.

| | | | | | | | | | | | | | | | |
|---|---|---|---|---|---|---|---|---|---|---|---|---|---|---|---|
| Key: | G | D | C | B | B | W | K | N | O | C | T | R | B | A | C |
| Plain: | S | E | N | D | U | S | A | L | Y | S | A | N | D | E | R |
| Cipher: | z | u | w | i | x | u | s | n | d | r | d | d | i | q | s |

| A | B | C | D | E | F | G | H | I | J | K | L | M | N | O | P | Q | R | S | T | U | V | W | X | Y | Z |
|---|---|---|---|---|---|---|---|---|---|---|---|---|---|---|---|---|---|---|---|---|---|---|---|---|---|
| A o | A e | A m | A h | A a | A q | A i | A c | A l | A w | A s | A b | A j | A y | A t | A z | A v | A f | A k | A d | A g | A n | A p | A x | A u | A r |
| B m | B o | B l | B s | B i | B f | B r | B z | B u | B x | B k | B a | B q | B t | B h | B w | B e | B y | B j | B n | B d | B v | B b | B c | B g | B p |
| C w | C s | C u | C f | C x | C p | C t | C a | C o | C j | C v | C y | C k | C i | C q | C n | C r | C z | C m | C l | C h | C g | C e | C b | C d | C c |
| D h | D i | D f | D r | D k | D x | D q | D o | D v | D s | D m | D p | D l | D u | D y | D j | D w | D n | D t | D a | D b | D z | D g | D e | D c | D d |
| E q | E a | E y | E u | E n | E o | E j | E t | E p | E f | E l | E i | E g | E s | E m | E x | E b | E h | E r | E k | E e | E w | E c | E d | E v | E z |
| F t | F q | F d | F c | F w | F b | F u | F h | F y | F e | F j | F o | F p | F x | F s | F k | F z | F a | F i | F m | F v | F r | F f | F g | F l | F n |
| G y | G m | G t | G q | G o | G u | G n | G x | G s | G v | G p | G w | G e | G k | G r | G l | G i | G j | G z | G h | G a | G c | G d | G f | G b | G g |
| H d | H l | H j | H a | H m | H z | H y | H f | H x | H u | H o | H n | H w | H r | H b | H v | H q | H e | H s | H g | H c | H p | H k | H i | H t | H h |
| I r | I d | I x | I t | I b | I n | I a | I l | I q | I p | I u | I e | I o | I c | I v | I s | I g | I w | I f | I j | I z | I m | I y | I h | I k | I i |
| J p | J t | J h | J z | J s | J w | J e | J r | J m | J c | J f | J q | J a | J v | J x | J d | J y | J g | J b | J i | J u | J o | J l | J n | J j | J k |
| K v | K n | K p | K m | K d | K s | K o | K y | K z | K q | K b | K u | K c | K g | K w | K f | K x | K r | K a | K e | K t | K l | K h | K k | K i | K j |
| L s | L h | L b | L y | L r | L v | L p | L i | L a | L t | L e | L x | L d | L n | L o | L g | L u | L q | L w | L c | L m | L k | L j | L z | L f | L l |
| M b | M g | M a | M k | M h | M y | M v | M s | M j | M z | M d | M r | M t | M q | M e | M o | M n | M x | M c | M f | M l | M i | M w | M u | M p | M m |
| N u | N k | N w | N v | N e | N i | N g | N q | N t | N y | N x | N h | N r | N l | N z | N c | N m | N d | N p | N b | N s | N a | N o | N j | N n | N f |
| O a | O b | O z | O w | O g | O e | O k | O d | O c | O r | O h | O f | O i | O p | O l | O m | O s | O v | O y | O q | O o | O j | O n | O t | O x | O u |
| P j | P w | P k | P x | P p | P c | P l | P v | P e | P i | P g | P d | P f | P o | P u | P y | P t | P s | P n | P z | P r | P h | P a | P q | P m | P b |
| Q e | Q f | Q v | Q g | Q z | Q a | Q d | Q n | Q i | Q k | Q y | Q j | Q b | Q m | Q c | Q t | Q h | Q l | Q u | Q o | Q w | Q x | Q r | Q p | Q s | Q q |
| R i | R v | R s | R d | R l | R t | R b | R j | R w | R o | R z | R m | R n | R h | R g | R u | R c | R k | R e | R x | R p | R f | R q | R r | R y | R a |
| S l | S c | S r | S b | S j | S k | S z | S m | S g | S d | S a | S v | S x | S e | S f | S i | S o | S p | S h | S w | S n | S y | S u | S s | S q | S t |
| T f | T j | T g | T i | T v | T r | T c | T e | T n | T l | T w | T z | T m | T b | T a | T q | T p | T u | T d | T y | T k | T t | T x | T o | T h | T s |
| U n | U x | U c | U e | U y | U g | U f | U w | U b | U h | U i | U k | U z | U d | U p | U r | U l | U t | U q | U v | U j | U u | U s | U m | U a | U o |
| V k | V r | V q | V n | V t | V l | V m | V p | V d | V g | V c | V s | V y | V j | V i | V h | V a | V o | V x | V u | V f | V b | V z | V w | V e | V v |
| W c | W p | W n | W o | W f | W j | W x | W u | W r | W a | W t | W g | W h | W z | W k | W b | W d | W i | W l | W s | W q | W e | W m | W v | W w | W y |
| X z | X u | X i | X p | X c | X d | X w | X g | X h | X b | X n | X l | X s | X f | X j | X e | X k | X m | X v | X r | X y | X q | X t | X a | X o | X x |
| Y g | Y z | Y e | Y l | Y u | Y m | Y h | Y k | Y f | Y n | Y q | Y c | Y v | Y a | Y d | Y p | Y j | Y b | Y o | Y t | Y x | Y s | Y i | Y y | Y r | Y w |
| Z x | Z y | Z o | Z j | Z q | Z h | Z s | Z b | Z k | Z m | Z r | Z t | Z u | Z w | Z n | Z a | Z f | Z c | Z g | Z p | Z i | Z d | Z v | Z l | Z z | Z e |

Deciphering is accomplished by operating inversely. The key is still written along the upper line, but this time the ciphertext is written directly under it. The plaintext will be found in the table, next to the cipher letter in the column of the corresponding key.

Once the key is destroyed, a cryptogram intercepted by the enemy will be impossible to solve, even if he obtains the table. For a cryptogram in this system could mean anything, depending on the key used. Therefore, with the same tableau but a different key, the cryptogram

| | | | | | | | | | | | | | | | |
|---|---|---|---|---|---|---|---|---|---|---|---|---|---|---|---|
| Key: | G | D | C | B | B | W | K | N | O | C | T | R | B | A | C |
| Cipher: | z | u | w | i | x | u | s | n | d | r | d | d | i | q | s |
| Plain: | S | E | N | D | U | S | A | L | Y | S | A | N | D | E | R |

could stand for "Paul has fled, stop," "Rendezvous Paris," or any other 15-letter message.

As an example, let's use the same cryptogram and tableau, but a different key:

| | | | | | | | | | | | | | | | |
|---|---|---|---|---|---|---|---|---|---|---|---|---|---|---|---|
| New key: | K | D | C | B | P | M | L | W | N | C | L | T | A | I | X |
| Cipher (identical to previous text): | z | u | w | i | x | u | s | n | d | r | d | d | i | q | s |
| Plain: | R | E | N | D | E | Z | V | O | U | S | P | A | R | I | S |

concealed in a personal package, such as a pack of cigarettes.

If new keys didn't arrive on time, or if for one reason or another the agent was stripped of his key groups, he could use a memorized poem of which the home station had a copy. This was only an entirely exceptional emergency key, the agent using as many letters from the poem as there were to be enciphered, and never using them again.

The essential precaution to be taken is to cut off from the microfilm the key groups that were used to encipher the message and to destroy these immediately. These groups are printed on nitrated rice paper that catches fire instantly, dissolves in hot liquids, is edible, and can be swallowed without harm!

Each agent had his own personal table, printed on a 9- by 4½-inch silk handkerchief, and 10 microfilm pages, numbered 1 to 10: 5 for deciphering, marked "Home Station to Outstation"; 5 for enciphering, marked "Outstation to Home Station." Each page measured 4¼ by 4 inches and had 1,000 groups that could only be read with a powerful magnifying glass.

Adopted by the I.S. and S.O.E. during the war, this system is now virtually universal for spies. It is the only completely inviolable procedure, whether used manually or electronically. Nevertheless, it has one inconvenience: the production and distribution of the series of random letters. The key has to be the same length as the message to be enciphered and can only be used once. This is why the system is hard to use at stations that have a large number of groups to transmit and that must therefore have enough key in their safes for their traffic.

For secret communications, the one-time pad, with the Vigenère table, was the best available.

When the agent had used up his keys, the home station dropped a new 10-page set by parachute. The pages were

### THE SYSTEM FOR INTERALLIED MISSIONS (JUNE THROUGH OCTOBER 1944)

#### Code Lists Enciphered by the One-Time Pad

Because of the considerable increase in clandestine radio communications, and to cut down on the length of transmissions, the interallied missions that were parachuted behind German lines, usually in uniform, were supplied with a large silk handkerchief (approximately 20 inches square) on which were printed both the expedition's table and a list of 4-letter code groups. Each group stood for a complete phrase.

On the handkerchief were more than 600 code groups for standard phrases. The one-time pad sheets were delivered separately. The code groups and phrases were printed in alphabetical order so that a single handkerchief could be used for both en- and deciphering. Here are some examples:

All containers received safely. Stop.: AOAO
Ambushed and killed enemy group at . . . : AUAU
Assistance urgently needed in area of . . . : BNBN

German division advancing in direction of . . . : CGCG
Militia active in area of . . . : LNLN
Safe house found for you in area . . . : PYPY
Shot up many and set fire to train . . . : QPQP
Staff officers liquidated and headquarters shot up . . . : RDRD
Telecommunications attacked and destroyed between . . . and . . . : RXRX

A short message of 6 or 7 groups was all that was needed to give notice that all containers had arrived safely, that a German division was advancing in the direction of such and such a town, or that attack planes were needed for support between such and such a point. The entire transmission required no more than 15 seconds.

The reader will certainly be intrigued by the repetition of the bigram within each code group (AOAO, CGCG, etc.). This seems illogical since a single bigram would be sufficient and 2 letters instead of 4 would cut down on transmission time. Actually, the repetition was an excellent way to cut down on coding errors and facilitate the decoders' work in those instances when an error could have slipped by unnoticed.

The phrases on these code handkerchiefs indicate a change of tactics. By the time they were used, open fighting had broken out. Instead of the groups with the dark meanings which characterized the clandestine operations of secret agents during previous years, the vocabulary is now slanted a lot more toward commando operations.

One can imagine, in reading these groups, the extraordinary teams of the Special Air Service, staging raids far behind the lines of a German army in full retreat. The S.O.E. was not equipped to carry out these actions. Subversive clandestine operations in France no longer had any reason to continue, and the networks had died out by the end of October 1944. Only the I.S. networks and the stations remained.

The secret home station 53-C, which had opened on April 4, 1944, ceased operations on November 3, 1944. The week of October 6, the 53-A home station had both transmitted and received only 10,000 groups as opposed to 180,000 during the week of July 19. The 53-B home station still had considerable traffic, averaging 80,000 groups a week. However, the majority of these communications were destined for networks operating in European countries which were still occupied.

## Internal Communications of a Net

### Invisible Inks

The S.O.E. employed invisible writing achieved by what is usually called "sympathetic ink."

Several recipes were taught in the secret S.O.E. school at Beaulieu in southern England. All had a base of simple ingredients that were easy to get in occupied France. Most were composed with nonprescription drugs issued for mild ailments, such as:

- One part bran mash diluted in 5 parts water. The message was developed with a cotton ball soaked in any depilatory solution that contained sulfur.
- A half-tablet of phenolphthalein laxative, 2 teaspoons of ammonia, and 2 soup spoons of water. The message was developed with a solution of common soda in water.

A very simple formula for ink consisted of 1 part alum to 100 parts water. Development was achieved by passing a hot iron over the sheet of paper. During this time of extreme poverty, many country families tanned rabbit skins, and the purchase of alum, which was necessary for tanning, did not arouse the least suspicion.

### The Hidden Message or "Barn Code" System

A fairly complex, but perfectly efficient, communication system was to hide a short message in an ordinary letter. This was known as the "Barn Code." Since this was such an important system, we will give a detailed example. Let's assume that the agent received the following letter in the mail:

Dear François,

How are you? I'm sorry I haven't been more effective in answering my mail, but you know how it is. Anyhow, I'm writing now to let you know that if you want to get some tires, you'd better get out here fast. This is the story:

Perhaps you will remember Pierre Bossuet. He was very taken with Suzanne, though nothing came of it. Anyhow, I met him in the Café des Artistes this morning and he told me that, coming from Paris, quite close to his place his old Citroën finally totally broke down. It's finished, but the tires are good. I said they would be a great help. He agreed, and tomorrow I'll send Annemarie to pick them up. You will have to move quickly—they're almost the same as weapons in the struggle for food these days. And they help us to find all kinds of supplies as well. Speaking of that, why don't you two come down here to the country for a weekend? We'd fatten you up! It would be marvelous to see you both again. So let me know if and when it's possible.

Cordially,
Jean

P.S. Speaking of food, perhaps the two of you could use some dried beans we have left. I think that Georges is going to Paris in two weeks, and he can probably bring you a kilo.

There is no reason for the attention of even the most mistrustful of censors to be drawn to this letter! However, . . .

Extract the tenth word of the first paragraph ("effective") and use it as a key, numbering its letters alphabetically to create a numerical key. Then write the second paragraph of the letter under this key.

The secret text is read by taking the word under column 1 in row 1, the word under column 2 in row 2, and so on: "Pierre taken in Paris. Close down A. Send quickly weapons, food, supplies."

Words that are too revealing are represented by the letter-word A—this message has only one. They are spelled out, enciphered in Playfair, in the postscript.

By convention, this cryptogram begins with the first letter of the fourth word of the postscript and continues as the first letter of every fourth word, as follows: PERHAPS, YOU, DRIED, LEFT, GEORGES, PARIS, AND, BRING.

Therefore the ciphertext would be: PY DL GP AB. If it had been previously agreed that the Playfair square would be made up with the key words SHAKESPEARE AND MILTON (though this time in the French way, suppressing the W, which would be replaced in a plaintext by two Vs), we would have:

| | | | | |
|---|---|---|---|---|
| S | H | A | K | E |
| P | R | N | D | M |
| I | L | T | O | B |
| C | F | G | J | Q |
| U | V | X | Y | Z |

| | | | | |
|---|---|---|---|---|
| Cipher: | PY | DL | GP | AB |
| Plain: | *du* | *ro* | *cn* | *et* |

The entire deciphered text would read: *Pierre taken in Paris. Close down Duroc net. Send quickly weapons, food, supplies.*

Although this system is long and it requires a lot of effort to draw up a plausible letter, it is practically impervious to analysis provided that all revealing words or words that normally would not be found in family or friendly correspondence are replaced by the letter-word A one after another and hidden in the postscript.

This system could only be used to pass short and urgent messages, of course. The use of a spontaneous and familiar tone, like that in a letter dashed off on the spur of the moment, was recommended.

We will not discuss microfilm and microdots here. These techniques, well-known nowadays, were used by the S.O.E. But properly speaking, they are not an extension of cryptography. They are simply an easy method for reducing documents or records. During wartime, the transmission of correspondence by microdot presents considerable problems. If there is a search, the equip-

| E<br>2 | F<br>5 | F<br>6 | E<br>3 | C<br>1 | T<br>8 | I<br>7 | V<br>9 | E<br>4 |
|---|---|---|---|---|---|---|---|---|
| Perhaps | you | will | remember | **PIERRE** | Bossuet. | He | was | very |
| **TAKEN** | with | Suzanne, | though | nothing | came | of | it. | Anyhow, |
| I | met | him | **IN** | the | Café | des | Artistes | this |
| morning | and | he | told | me | that, | coming | from | **PARIS**, |
| quite | **CLOSE** | to | his | place | his | old | Citroën | finally |
| totally | broke | **DOWN**. | It's | finished | but | the | tires | are |
| good. | I | said | they | would | be | **A** | great | help. |
| He | agreed | and | tomorrow | I'll | **SEND** | Annemarie | to | pick |
| them | up. | You | will | have | to | move | **QUICKLY**— | they're |
| almost | the | same | as | **WEAPONS** | in | the | struggle | for |
| **FOOD** | these | days. | And | they | help | us | to | find |
| all | kinds | of | **SUPPLIES** | as | well. | Speaking | of | that, |
| why | don't | you | two | come | down | here | to | the |
| country | for | a | weekend? | We'd | fatten | you | up! | It |
| would | be | marvelous | to | see | you | both | again. | So |
| let | me | know | if | and | when | it's | possible. | |

ment for reading and producing the microdots becomes damaging evidence.

## Radio Communications Among S.O.E. Staffs

There remain only systems employed at the highest level, in other words, for radio communications between different S.O.E. staffs. The headquarters might be Force 136, between Calcutta and London, or the forward command post at K'un-Ming in China.

Properly speaking, we are no longer dealing with clandestine communications, since the staffs carried out their operations in the homeland or in Allied territories not controlled by the enemy. The classic cryptographic procedures of large army units were consequently employed. However, absolute security was required that could only be given by the one-time pad, excluding the use of any cryptographic machines. Therefore, the thick code books of diplomacy had to be used again. Stored in safes, these were made up of 4-digit groups (10,000 possible groups) which were superenciphered by adding (without carrying) key groups from the one-time pad, which were destroyed as they were used up.

A part of such a code might have looked like this:

| Encoding section | | | | | Decoding section | | |
|---|---|---|---|---|---|---|---|
| *M* | : | 4631 | 3254 | 4681 | 3254 | : | *M* |
| *Mac* | : | 5362 | | | 3255 | : | *Relief* |
| *Macadam* | : | 4316 | | | 3256 | : | *Pocket* |
| *Machine* | : | 3214 | 8126 | 3142 | 3257 | : | *Outbreak* |
| *Marvel* / *Mary* | } | 4782 | | | 3258 | : | *Midship* |
| | | | | | 3259 | : | *Michael Relinquish* |

Pads of 4-digit groups were used for superenciphering. As soon as the message was superenciphered, the key groups were cut off and burned along with the rough draft of the message. The ashes were carefully crushed. For easier radio communication, messages were sent out in 5-digit groups.

Deciphering was carried out in reverse order, in other words, by subtracting, without borrowing, the key groups from the cryptogram groups. In this way, the code groups whose meanings appeared in the codebook could be found.

It should be noted that widely used terms could have 2 or 3 different representations, as does the letter *m* or the word *machine* in the table above. Less common words, such as *marvel* and *Mary*, could share the same code group as long as they did not create any ambiguity. Such a group is called a "polyphone."

There is an anecdote about these 2 words. One morning in 1945, a few months before Japan's surrender, the S.O.E. base at K'un-Ming received a telegram from London announcing the imminent arrival of 4782, which the code room translated as *the arrival of a marvel*. When a young cipher clerk named Mary got off the announced plane, the code team insisted it hadn't made a wrong interpretation of the message after all, and from that point on, Mary was never known by any name other than "The Marvel."

This system offered complete security, of course. The only solution for the Japanese would have been for their agents to secretly photograph the code list and the key groups. Probably every precaution had been taken, and it was highly unlikely that this happened.

By contrast, in 1945 the S.O.E. was still using double transposition for communications between the home station and the networks in the Far East. Still, it appears that the Japanese cryptanalytic services were far from able to break the messages of the Westerners that were enciphered in this manner.

## CONCLUSION

Cryptography is the twin sister of writing. Appealing to the deeper instincts of man, it has always been the object of intense and scholarly research. During the 3,000 years of its existence, cryptography has undergone 3 successive mutations. The first was during the Renaissance. This was when all the theories were established and the transposition and substitution systems that govern classic cryptography were invented.

The second mutation came about through an invention which upset the principle of the secret missive: radio, which submitted its broadcasts to all areas and all listeners. World War I used it, and the initial years of the war saw eminent mathematicians solve all the existing systems. A few hours were needed to solve the easiest, a few weeks, the hardest.

They were foiled by machines invented during and several years after the war that resisted the efforts of the best specialists. These machines, remember, automated a polyalphabetic substitution system with a very long key.

The last mutation took place during the first part of World War II. Long delays by even the best mathematician-cryptanalysts were cut down considerably by using tabulators. The last parry was the adoption of the one-time pad system, or the infinite incoherent key.

This defense, however, was not used universally since it required the preparation, distribution, and storage of a volume of keys equal to the volume of messages to be enciphered. This was impractical for heavy traffic. The second solution was through machines capable of creating longer and longer incoherent sequences, although the secrets they protected could only exist for a certain length of time.

A turning point for cryptography, the Second World War logically saw a fairly large number of systems succeed one another. The S.O.E. provided a typical example, first by purely and simply reviving the British army's field cipher from the previous war: the Playfair. Later, probably because of the intervention of the cipher experts, there came the decision to abandon the Playfair, which was no longer safe enough for important messages sent by radio, and its replacement by double transposition. Without a doubt the choice was unfortunate for more than one reason. But, at that time, the S.O.E. wasn't the only organization using this system, which could seduce a cryptologist in his study but was certainly detested by an agent in the field.

Double transposition keys appeared successively in the form of a book, of a line of verse, and prepared keys (the A–Z cipher). Each phase brought simplification and increased security. In addition, when the A–Z cipher appeared, it was decided that the agent would no longer be required to learn anything by heart or to accept the risk of receiving something that, in case of capture, could convict him.

At the end of 1942, on the highly authoritative advice of Intelligence Service professionals, it was decided to abandon all transposition systems in favor of the one-time pad. Since an immediate implementation would have required a recall of all missions, it was decided to first introduce systems superior to double transposition to increase the security of certain networks. The Delastelle system, whose high level of security results from the fact that it is both a transposition and a substitution, is the most notable among these. In addition, this system did not require long key lists.

The numerical system, with its incoherent, nonrepeating key, guaranteed absolute communications secrecy to agents for the first time, although under a form that was still too complicated.

Finally, starting in August 1943, all F networks employed the one-time pad, along with the Vigenère tableau, to the satisfaction of everyone. The only problem remaining was that of supplying the agents with key lists.

It can therefore be said that at least until the end of 1942 (and the summer of 1943 for some networks), the Germans were technically capable of deciphering S.O.E. radiograms. After these dates, they were in the dark, an absolute and total darkness.

By contrast, the French secret service continued to use the double transposition system right up until the end of hostilities in France. It is to be hoped that this did not have any unfortunate consequences. Nevertheless, the imprudence was great.

Chapter V

# Tangible Links: Black Squadrons and Secret Navigation

## DELIVERY MISSIONS AND PICKUP MISSIONS

### Armstrong Whitworth "Whitley" Bomber Mark V (1939)

***Bottom View***

*Scale: 1/150*

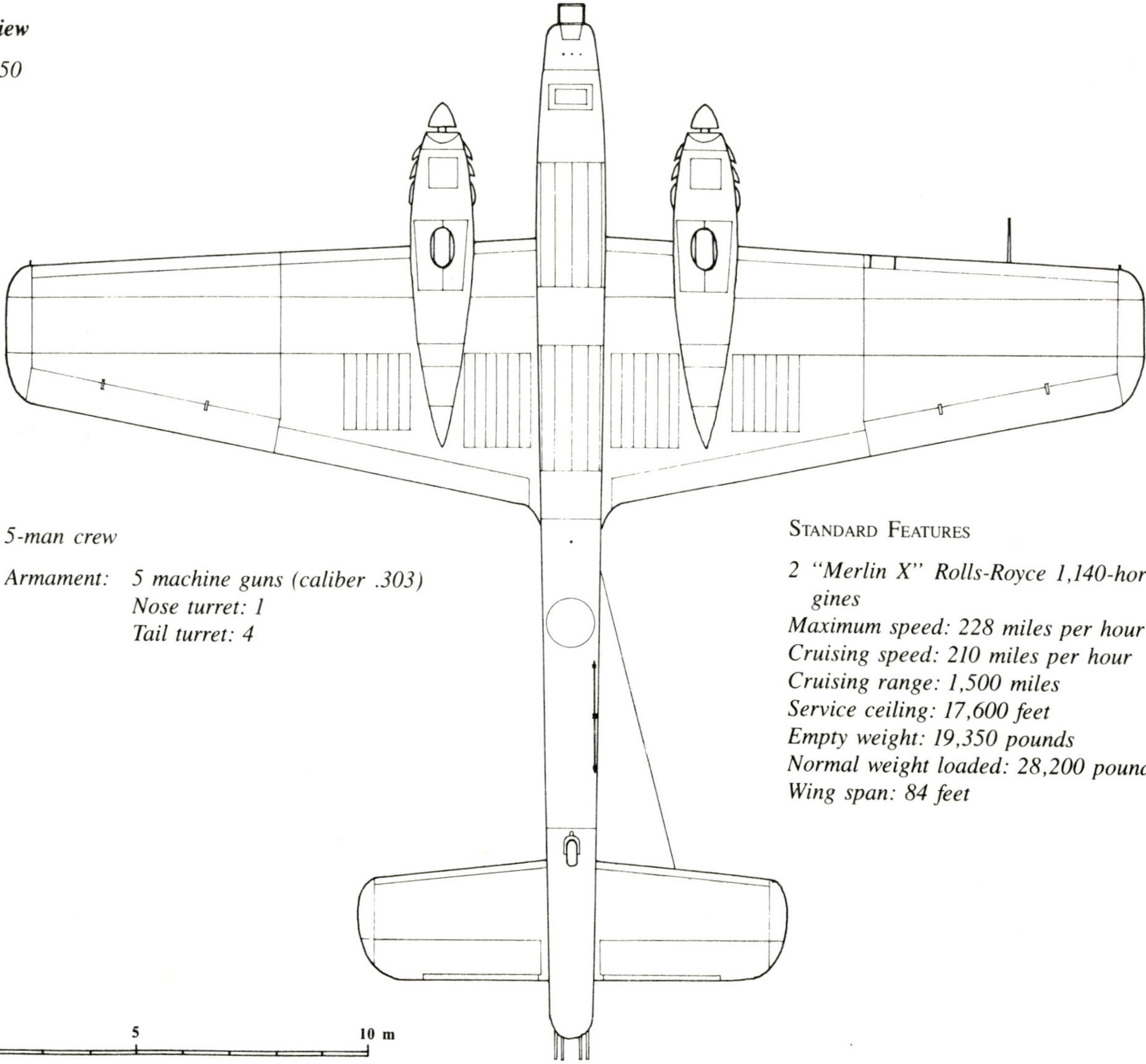

*5-man crew*

*Armament:* *5 machine guns (caliber .303)*
*Nose turret: 1*
*Tail turret: 4*

STANDARD FEATURES

*2 "Merlin X" Rolls-Royce 1,140-horsepower engines*
*Maximum speed: 228 miles per hour*
*Cruising speed: 210 miles per hour*
*Cruising range: 1,500 miles*
*Service ceiling: 17,600 feet*
*Empty weight: 19,350 pounds*
*Normal weight loaded: 28,200 pounds*
*Wing span: 84 feet*

0 5 10 m

## *In Flight*

*Scale: 1/150*

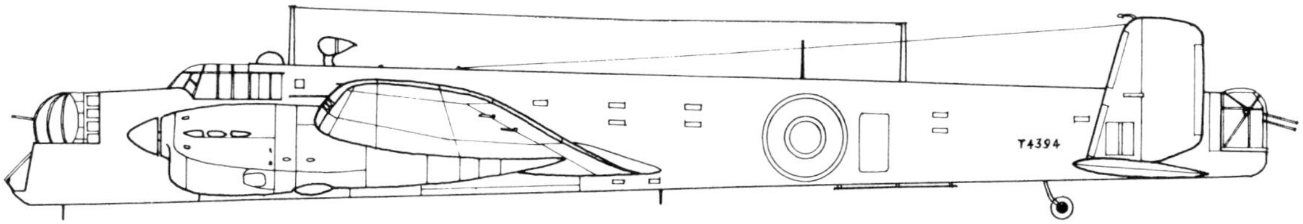




*A night bomber with a long radius of action, the Whitley utilized 1934 specifications to give the R.A.F. its first modern heavyweight bomber.*

*In 1936, the Whitley Mark I was powered with two 795-horsepower engines. Other models quickly followed, each with increased engine power and improved performance. The 1938 Mark IV had two 1,030-horsepower engines and was the first bomber to have four .303 machine guns in a tail turret.*

*However, by 1939 the Whitley seemed outclassed, but more than that, outmoded. Its armament, speed, ceiling, and useful load were insufficient. The old Whitley did have two advantages. With an auxiliary fuel supply, it had a greater cruising range than any other plane in service. Used by Bomber Command in 1940 and 1941, it was then handed over to Coastal Command, where it rendered numerous services in the antisubmarine campaign. The Whitley ended its career as a transport plane.*

*Meanwhile, the secret services needed a plane which could transport both food supplies and agents deep into the occupied territories. They chose the Whitley. The fact that this plane was outmoded mattered little for clandestine activities. An agreement was drawn up with Bomber Command to supply 2, then 4, then 10 Whitleys, finally peaking with 17 at the beginning of 1942.*

*Used by the Intelligence Service starting in August 1940, and by the S.O.E. starting in March 1941, for nearly 2 years the old Whitley was the principal carrier for the first agents who infiltrated by parachute. Each month the Whitley brought these agents a load of sabotage material.*

*The plane had many archaic features such as angular lines and an excessively long and flat fuselage, which made it look like a "flying girder."*

*Its sinister appearance was accentuated by the airplane's being painted dull black to avoid reflecting searchlight beams and moonlight.*

*The bottom view of the plane was also disturbing. Because of its shape and its light weight in relation to its size, the Whitley resembled one of those hollow-boned herons which weigh next to nothing and are all feet and wings.*

# Vickers "Wellington" Bomber Mark III (1942)

***Bottom View***

*Scale: 1/150*

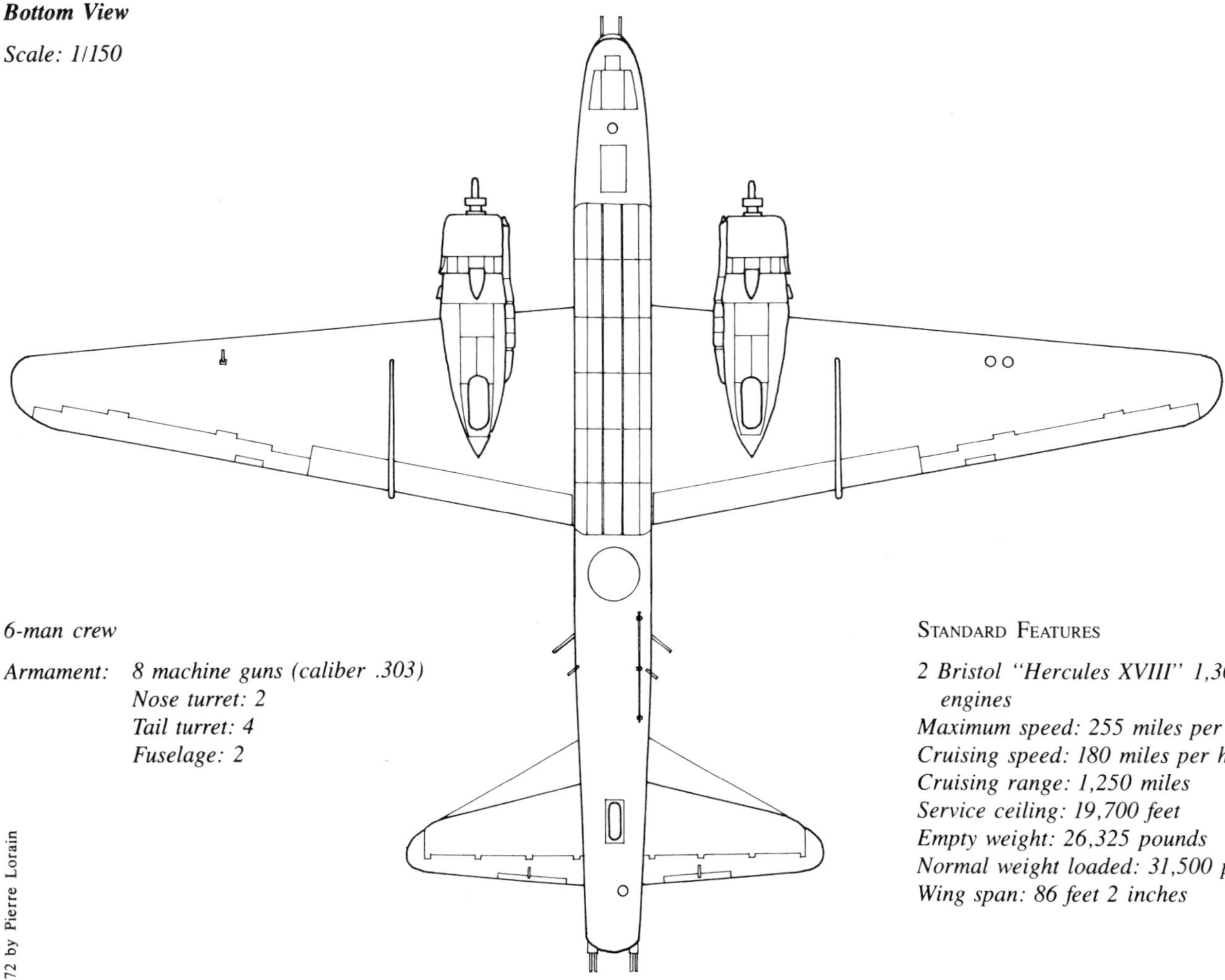

*6-man crew*

*Armament:* *8 machine guns (caliber .303)*
*Nose turret: 2*
*Tail turret: 4*
*Fuselage: 2*

Standard Features

*2 Bristol "Hercules XVIII" 1,300-horsepower engines*
*Maximum speed: 255 miles per hour*
*Cruising speed: 180 miles per hour*
*Cruising range: 1,250 miles*
*Service ceiling: 19,700 feet*
*Empty weight: 26,325 pounds*
*Normal weight loaded: 31,500 pounds*
*Wing span: 86 feet 2 inches*



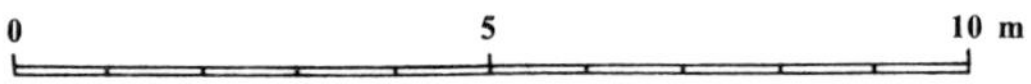

## *In Flight*

*Scale: 1/150*

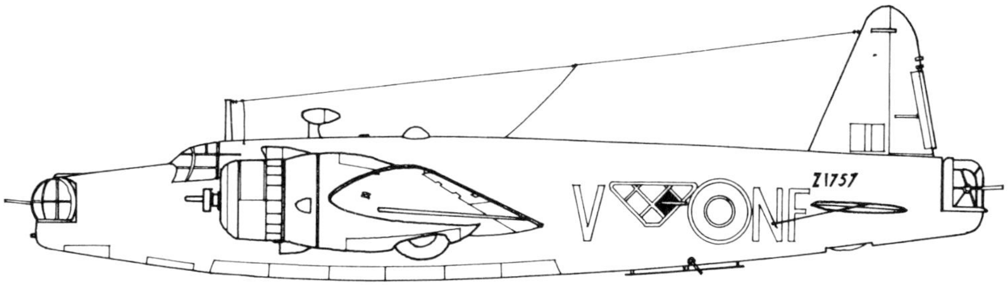




*Like the Whitley, the Wellington was a "first generation" plane since it was based on 1932 specifications. Only it was fortunate enough to be born under a lucky star. It was a definite success. Brilliantly designed, the Wellington had an alveolar structure, a kind of skeleton which, before it was covered over, gave the fuselage the appearance of a woven basket. This frame, which was rather unusual and very strong, enabled the Wellington to support increasingly powerful engines without harming its airframe. In 1939, the Mark I flew with two 1,000-horsepower engines. In 1945, the Mark X could hold two 1,600-horsepower engines.*

*The Wellington's pleasing features enabled it to handle many different tasks. In 1943 it was retired from Bomber Command and handed over to Coastal Command. It was also extensively used as a training plane for radar, radio operations, and night bomber crews. Finally, Transport Command used the Wellington a lot because of its capacious fuselage.*

*During the initial years of the war, it was impossible for the secret services to use the Wellington since the demand for the plane was so high. Nevertheless, Professor Michael R. D. Foot points out that 1 Wellington was in service in Squadron 138 and 2 in Squadron 161 in February 1942.*

*This plane was effectively employed, although in small numbers, by the special services. Except for the fact that it was used to train agents for 1944 parachute jumps, I.S. and S.O.E. records contain no other precise information about its war missions. There, as elsewhere, its performance was probably completely satisfactory during its 2½ years of service.*

*Machine guns were located on both sides of the fuselage. These were added after the disastrous Schillig Roads incident on December 18, 1939, in which 15 Wellingtons in a 22-plane formation were shot down by a German fighter squadron which attacked them broadside.*

*Note the elegance of line and the rounded, potbellied fuselage, unlike the skinny beamlike body of the Whitley. There was an I.F.F. (Identification Friend or Foe) aerial, which enabled the plane to slip into the middle of an English radar surveillance network without any danger. Above the cockpit was a pole to hold up the aerials, a directional loop in a protruding casing, and an astrodome so the navigator could fix his position with the sextant.*

# Handley-Page "Halifax" Heavy Bomber B Mark III (1943)

***Bottom View***

*Scale: 1/150*

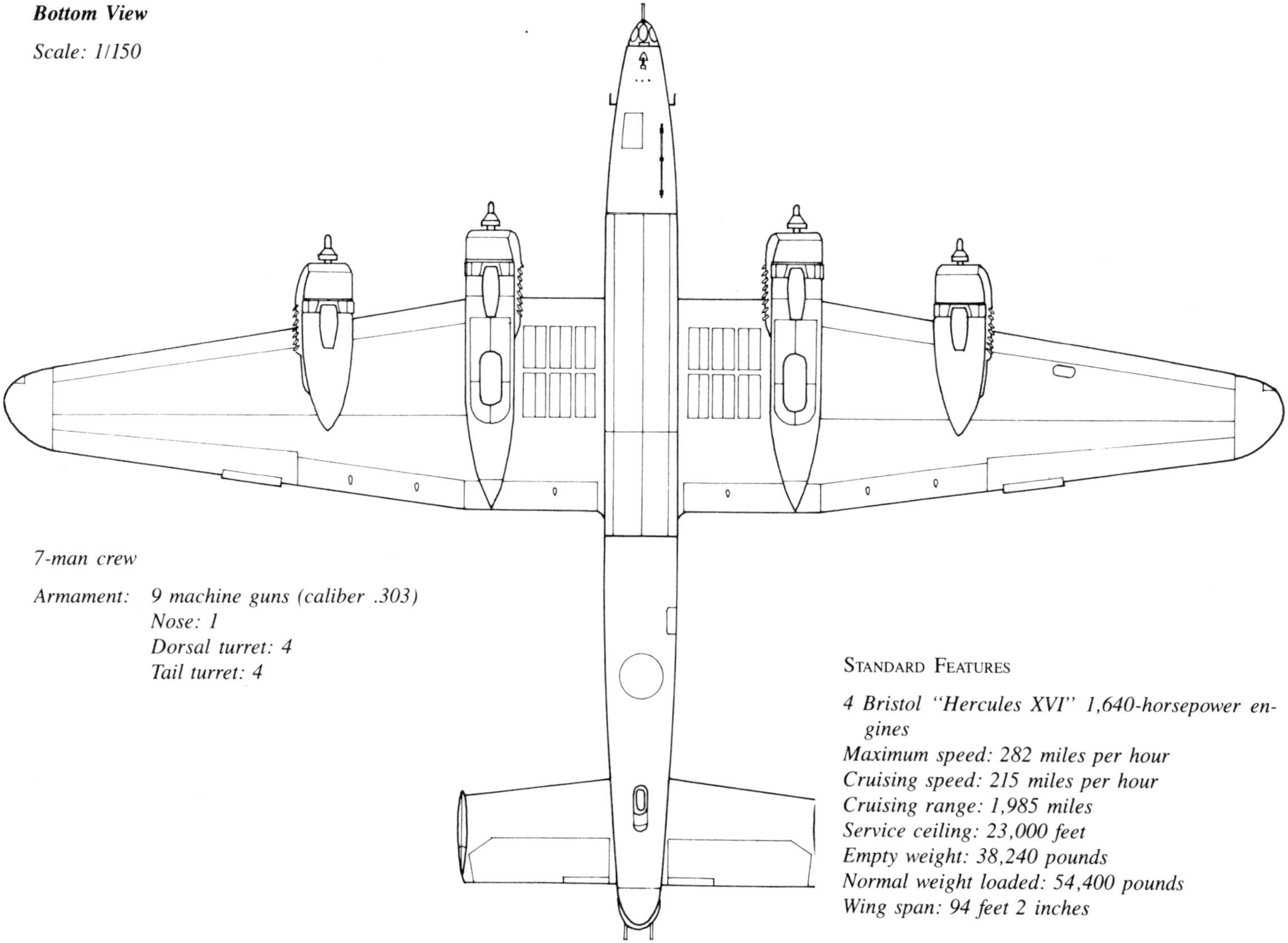

*7-man crew*

*Armament:* *9 machine guns (caliber .303)*
*Nose: 1*
*Dorsal turret: 4*
*Tail turret: 4*

Standard Features

*4 Bristol "Hercules XVI" 1,640-horsepower engines*
*Maximum speed: 282 miles per hour*
*Cruising speed: 215 miles per hour*
*Cruising range: 1,985 miles*
*Service ceiling: 23,000 feet*
*Empty weight: 38,240 pounds*
*Normal weight loaded: 54,400 pounds*
*Wing span: 94 feet 2 inches*

0 5 10 m

Note: *The 1942 Halifax B Marks were equipped with 4 liquid-cooled Rolls-Royce "Merlin XXII" 1,450-horsepower engines.*

## *In Flight*

*Scale: 1/150*

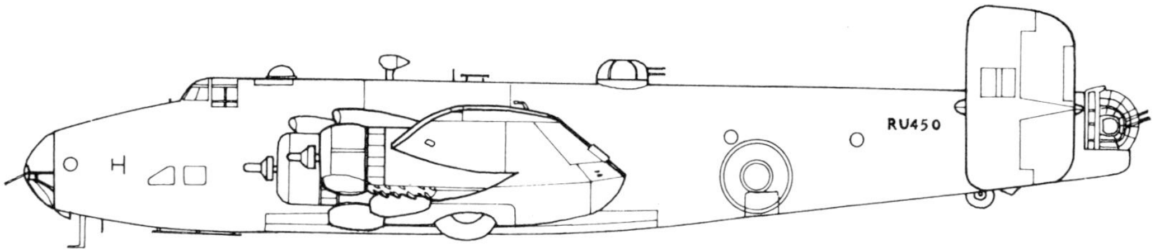




*After the Lancaster, the Halifax was the most celebrated of the heavy R.A.F. bombers. It belonged to the second generation of bombers, namely planes designed and built at the beginning of the war.*

*The Mark I first flew in October 1940 and had four 1,280-horsepower engines. The Mark VI, built in 1944, was equipped with 4 huge 1,800-horsepower engines.*

*Many persons were struck by the impression of power and speed that emanated from the Halifax silhouette.*

*Many aerials appeared on the fuselage: the I.F.F., S-Phone, Rebecca, different types of radar, V.H.F. (very high frequency) radio, and the directional loop. The Halifax had highly complex electronic equipment.*

*Three tiny aerial wires, spaced 6 inches apart, sprouted under the nose of the plane. These were the reception aerials for the S-Phone, which enabled the plane to converse on a secret frequency with the head of the "reception committee," awaiting a delivery somewhere in France. These antennas and this S-Phone guided the pilot into an invisible corridor and drew his plane like a magnet toward some far-off, out-of-the-way field.*

*Although these Halifax planes were valuable for Bomber Command, in August 1941 some were supplied to Squadrons 138 and 161 to relieve the old Whitley.*

*This relief coincided with a considerable increase in the number of S.O.E. networks. Furthermore, a new factor intervened, with the creation of the Compulsory Work Service for Germany. It was decided to create the foundations for a secret army with young insubordinates in this service. Therefore, it was advisable to train and arm them.*

*A substantial increase in tonnage to be dropped by parachute could be foreseen as a result of this new undertaking. Only the heavy, 4-engine planes could ensure such a service.*

*Between August 1941 and January 1944, or for 2½ years, some 20 Halifaxes worked for the secret services. If the Whitley was the first supplier of saboteurs, the Halifax was the supplier of small arms for the vanguard of a vast clandestine army.*

# Short "Stirling" Heavy Bomber B Mark III (1943)

***Bottom View***

*Scale: 1/150*

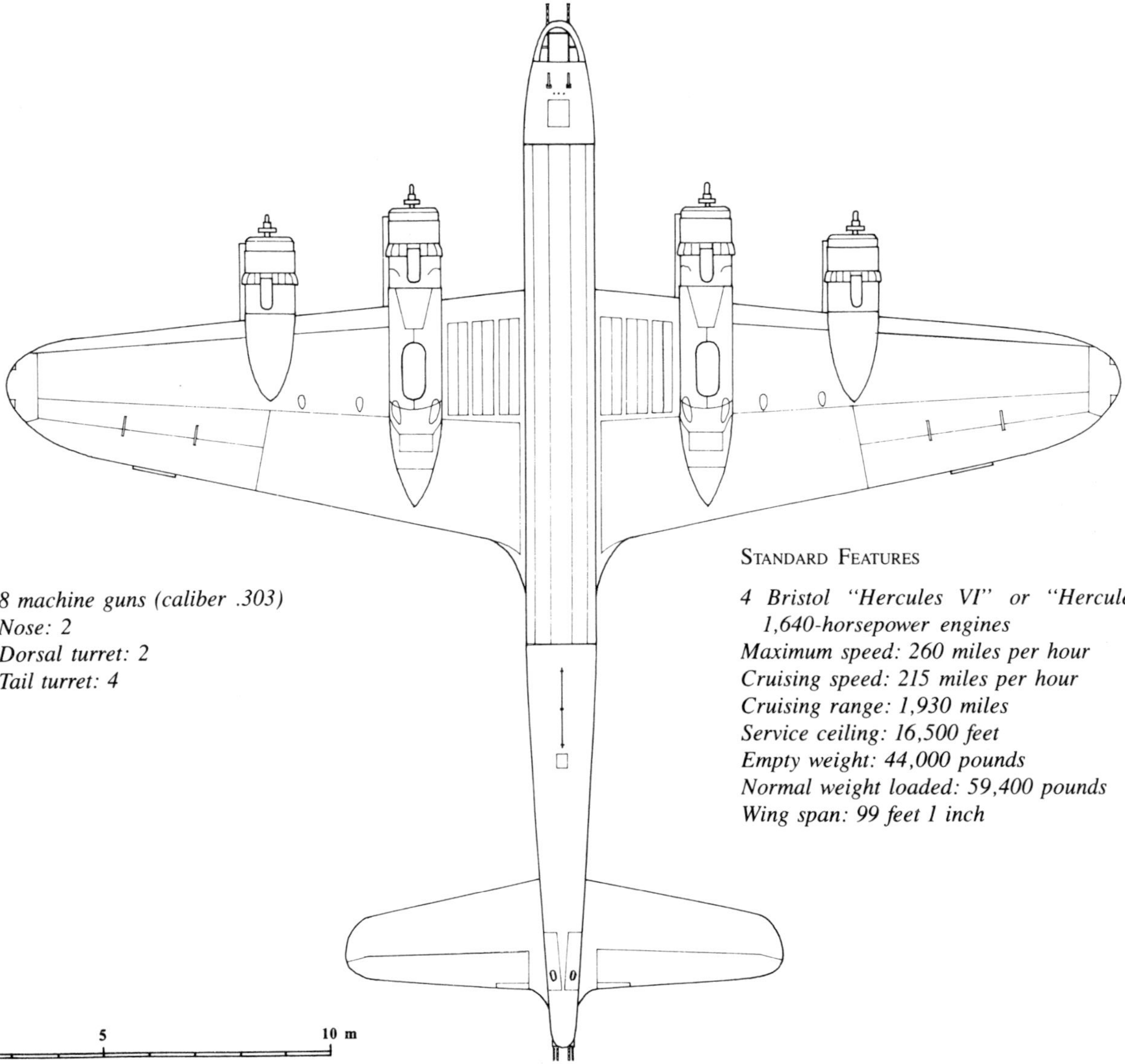

*8-man crew*

*Armament:* *8 machine guns (caliber .303)*
*Nose: 2*
*Dorsal turret: 2*
*Tail turret: 4*

Standard Features

*4 Bristol "Hercules VI" or "Hercules XVI" 1,640-horsepower engines*
*Maximum speed: 260 miles per hour*
*Cruising speed: 215 miles per hour*
*Cruising range: 1,930 miles*
*Service ceiling: 16,500 feet*
*Empty weight: 44,000 pounds*
*Normal weight loaded: 59,400 pounds*
*Wing span: 99 feet 1 inch*

0 5 10 m

## *In Flight*

*Scale: 1/150*

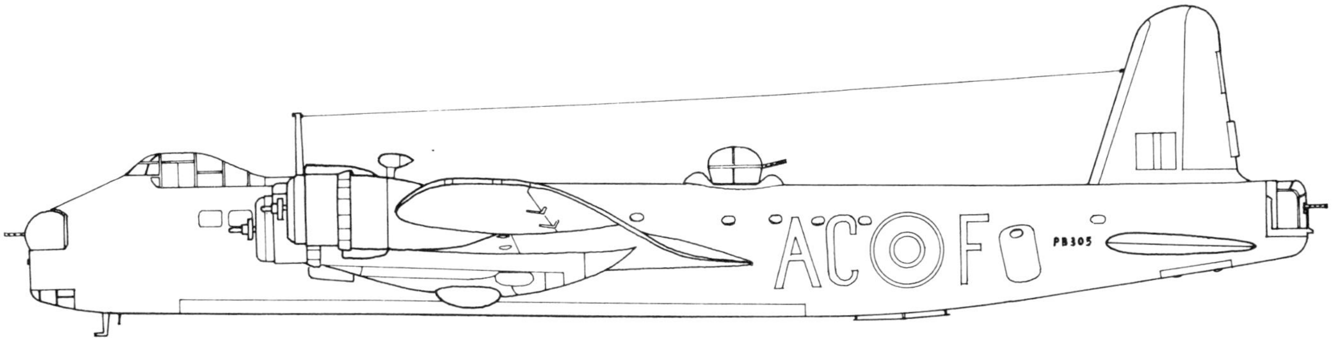


© 1972 by Pierre Lorain

*Like the Halifax, the Stirling belonged to the second generation of heavy bombers. It was the oldest of the 4-engine planes. The first Stirling took flight in August 1940, equipped with four 1,640-horsepower engines.*

*A Mark II model followed close behind. The Mark III was the last in the series. Bomber Command used this plane jointly with the Halifax, but, because of inadequate performance, it was dropped from service at the beginning of 1943.*

*Owing to its weight and enormous fuselage, the four 1,640-horsepower engines had a difficult time lifting the plane off the ground, and accidents at takeoff were numerous. For many reasons, the Stirling was a dangerous plane. On the other hand, once it was in the air, it could be easily maneuvered and, in spite of its excessive length, could escape a fighter unit or a too precise antiaircraft gun by executing swerves and sudden hops.*

*The leading edge of the wings was designed for cutting barrage balloon cables.*

*The general sturdiness of the plane and its useful load were exceptional.*

*On the other hand, its wings were too short in proportion to the fuselage, and its enormous weight prevented it from ever reaching an adequate ceiling, for German heavy antiaircraft constantly fired higher and higher, faster and faster, and more and more accurately.*

*The third-generation planes, the Lancasters or the American "Flying Fortresses," had ceilings of 25,000 and 35,000 feet, respectively. But these planes were intended from the first for bomber squadrons and were never handed over to Squadrons 138 and 161.*

*The Stirling's outline shows some archaic features which bring it somewhat closer to the Whitley—in particular, the flat beamlike structure of the fuselage and its disproportion in relation to the wings.*

*Between October 1943 and September 1944, or for 11 months, 22 Stirling bombers served as the last supply planes for the S.O.E. Most of their missions took place during the three months before the Normandy landing.*

*Large formations of B-17s from the United States 8th Air Force aided the Stirling with massive daytime drops in July 1944.*

# Westland "Lysander" Mark II (1939)

***Three-Quarters, Left Front View***

*(Landing Officer's View)*

*The plane is ready to touch down. The pilot, eyes fixed on the improvised ground lights, has just switched on his landing lights and is going to pass in front of the first ground light.*

*The Lysander, affectionately nicknamed "Lizzy," was a first-generation, fire-support and reconnaissance plane serving in the R.A.F. beginning in 1938. Combat exposed its faults: the plane was slow and heavy, and its armament was inadequate for ground support and for its own defense. Its outstanding features were sturdiness, maneuverability, and the ability to land and take off in a small field. For the secret services, this was the ideal plane. They lightened it, removed its armament, and fitted a detachable reserve tank and a metal ladder for rapid passenger boarding.*

*The Lysander set up regular service between England and France for 4 years.*

*The Lysander radio equipment in 1941 and 1942 was: a 5-tube, high-frequency TR-9 battery-operated radiotelephone (wave band of 5 megacycles); in 1943: a very high frequency radiotelephone, model TR-1143, with 4 pretuned channels in a wave band of 120 megacycles, operating on a 24-volt converter. In addition, there was a secret S-Phone or Rebecca system.*

*The aerial illustrated was for the earliest TR-9 set. The plane was completely painted in dull black.*

*In Flight*

*Scale: 1/150*

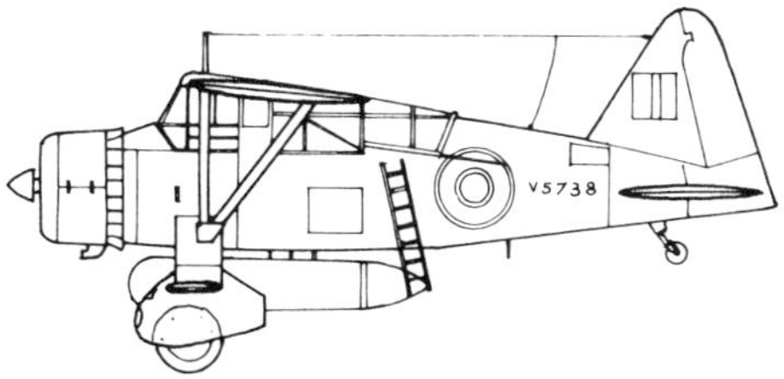


0 5 10 m

*Crew: 1 pilot*
*Passengers: usually 1 or 2, with room for 3 at the most, each with hand luggage*
*Empty weight: 2 tons*
*Normal weight loaded: 2.9 tons*
*Wing span: 48 feet*

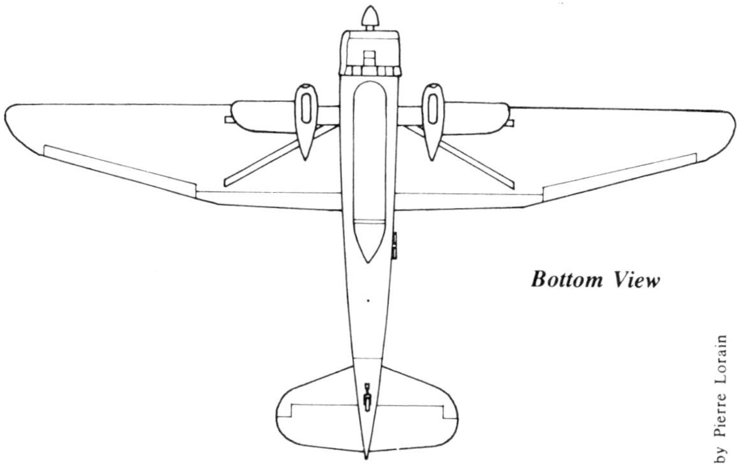

***Bottom View***

STANDARD FEATURES

*1 Bristol Perseus XII 900-horsepower engine*
*Maximum speed: 210 miles per hour*
*Cruising speed: 165 miles per hour*
*Cruising range: without reserve tank: 570 miles*
*with reserve tank: 900 miles*
*Service ceiling: 26,000 feet*

*Wind direction*

## Pickup Operation



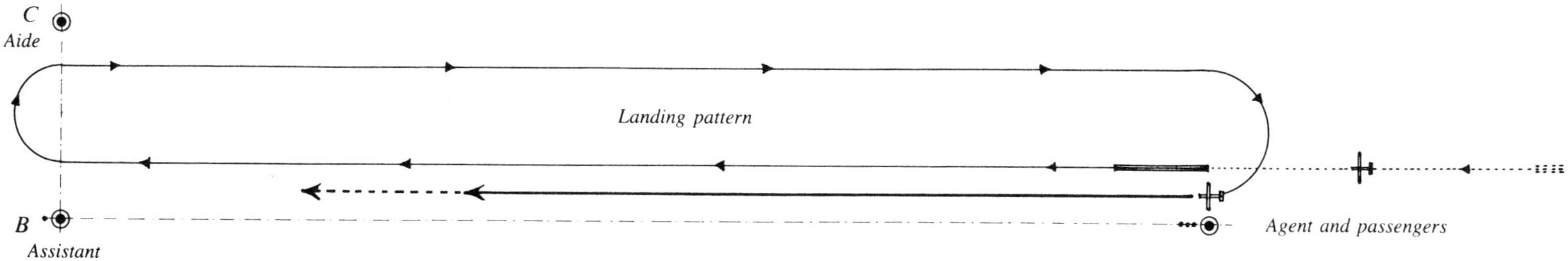


### *Landing and Takeoff Pattern, Field Beacons*

*The pilot flies into the wind along the axis of the airstrip at an altitude of 300 feet. He touches ground in front of light* A, *or no more than 200 feet beyond it. He brakes and turns between lights* B *and* C, *heads back toward his landing point, again faces the wind, and stops to the right of light* A.

0 50 100 m

NOTE: *If possible, the landing axis* A-B *is indicated by 3 or 4 flashlights placed in a line every 300 feet. A Eurêka ground beacon could advantageously be placed at* A.

# Lockheed "Hudson" Light Bomber Mark I (1939)

## *In Flight*

*Scale: 1/150*

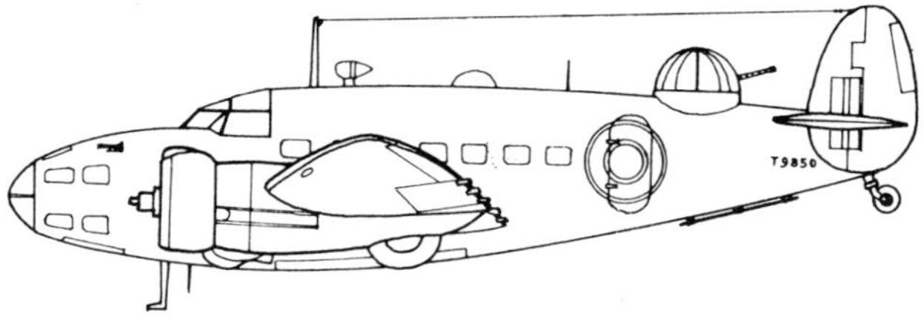


NOTE: *When these American planes arrived in England, the British equipped them with Boulton Paul dorsal turrets. From the beginning of 1942 until the end of operations in France, between 1 and 6 Hudsons were attached to Squadron 161 for pickup missions.*

*4-man crew*

*Armament: 4 machine guns*
*Nose: 2 fixed*
*Dorsal turret: 2*

***Bottom View***

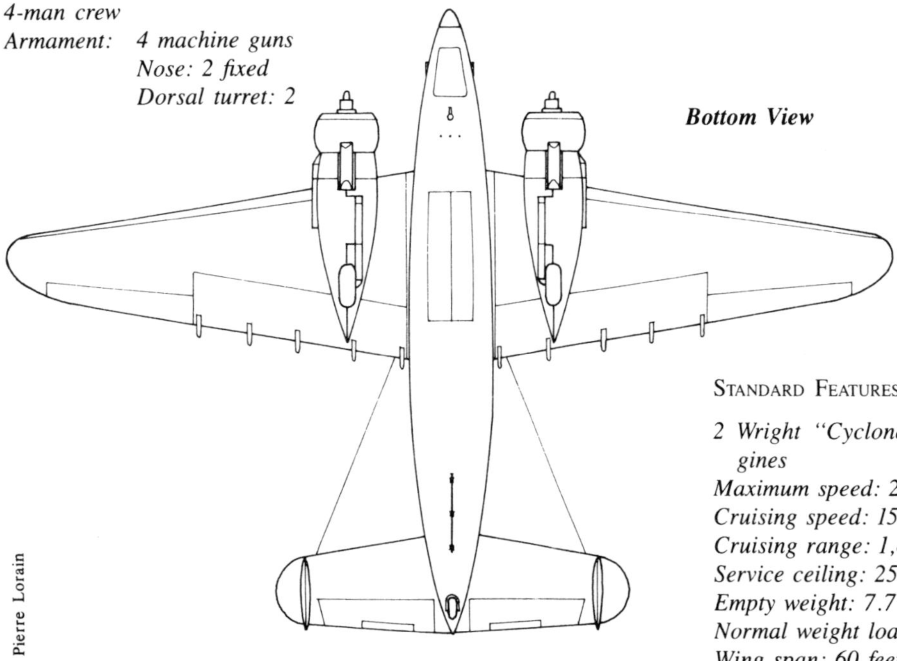

STANDARD FEATURES

*2 Wright "Cyclone" R. 1820, 900-horsepower engines*
*Maximum speed: 240 miles per hour*
*Cruising speed: 150 miles per hour*
*Cruising range: 1,800 miles*
*Service ceiling: 25,000 feet*
*Empty weight: 7.7 tons*
*Normal weight loaded: 12 tons*
*Wing span: 60 feet*
*The cabin could accommodate 12 passengers and a ton of cargo*



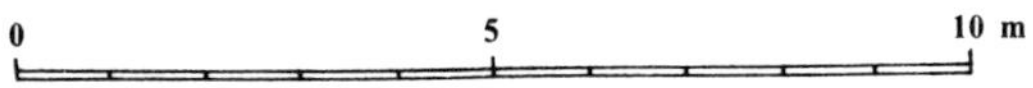


NOTE: *Hudson aircraft used toward the end of hostilities had 2 Pratt & Whitney "Twin Wasp" 1,200-horsepower engines.*

*After taxiing down the runway, the plane came to a stop, facing into the wind next to the agent's light. The engines did not stop turning, and the crew members stayed at their posts. Each member of the "reception committee" played a very precise role. Underground fighters guarded the area. At the left, the man kneeling is disassembling the secret "Eurêka" ground beacon which has controlled the plane over the last 90 to 15 miles and has automatically guided it into the axis of the field. Once folded up, the Eurêka, its power supply, and its aerial were placed in a small carrying case that protected it from dirt.*

*The agent wearing a topcoat conceals the S-Phone, whose aerial is visible. With it, he can communicate with the pilot or any crew member for 20 to 5 minutes before the plane appears, depending on its altitude. Likewise, after takeoff, and under the same conditions, the ground agent can give his final oral instructions to the departed agents. This instrument is virtually undetectable.*

# DEVICES FOR SECRET NAVIGATION

For clandestine pickup or drop operations, a direct link between the ground crew and the supply plane was practically indispensable. This kind of mission required instant radiotelephone communication in clear language.

Radiotelephones had been present in both civil and military aeronautics for a long time. What remained was to develop clandestine equipment with the following requirements: compactness, lightness, and confidentiality. The range would be limited to the optical horizon. An additional guarantee of security would be offered by the emission of a narrow beam, analogous to an invisible pencil of light. All these conditions were satisfied by using ultrashort waves, which were then also being used for radar.

On the other hand, an almost unsolvable problem for the navigator was that of locating his tiny target, at night, from a very low altitude, and, at times, in the fog.

The ideal solution consisted of miniaturizing a radio beacon that the reception committee would set up in the middle of the drop field. This would automatically guide the plane down a narrow corridor and indicate the exact spot where parachutes were to be released.

Starting in 1940, the British special services designed and manufactured this instrument. Special security precautions were taken to protect the secrecy of this instrument, which, until today, remained virtually unknown.

It consisted of the S-Phone and the Eurêka-Rebecca set.

The S-Phone fulfilled 3 functions at the same time: radiotelephone, homing beacon, and parachute drop spot indicator. A tiny instrument carried on a webbed harness, it had the following standard features:

Transmission frequency: 337 megacycles
Reception frequency: 380 megacycles
Power output: 0.1–0.2 watt
Equipped with 4 or 5 miniature tubes, depending on the model, in an olive green aluminum box
Dimensions: 7½ × 4 × 20 inches
Weight: Approximately 2¼ pounds
Total weight, including belt and batteries, came to about 15 pounds

S.O.E. agents used the S-Phone mostly for air-ground communications. But it was also routinely employed between the shore and a ship, a submarine, or a fast patrol boat during clandestine landing and removal operations, generally of I.S. agents.

The Eurêka device appeared during the second part of the war. It was a portable radar responder that fulfilled 2 functions: homing beacon and parachute drop spot indicator. Its range was greater than that of the S-Phone, but it could not be used as a radiotelephone.

Toward the end of the war, a secret mission was entrusted to S.O.E. agents in France for the benefit of the R.A.F.: Eurêka beacons with telescopic aerials were dropped with orders for them to be set up in specific geographical locations, most often in sheds or barns on farms. Started at nightfall and shut off every morning, these beacons constituted a clandestine homing device chain which guided R.A.F. bombers toward far-off strategical targets in Italy or Central Europe through secret routes far removed from antiaircraft zones.

# S-Phone Type 13/Mark IV (1943)

## *Ultrahigh Frequency Secret Radiotelephone*

*The set is held under the belt by 2 webbed straps. Once unfolded, the vertical dipole is plugged into its socket. The operator adjusts the frequency knob with his right hand until he gets the best reception possible.*

*The straps are attached to a belt holding 7 canvas pouches. Five of these hold 10 miniature rechargeable cadmium-nickel batteries; a sixth contains the vibrator power pack; and the seventh, the headset, microphone, and collapsible aerial when the set is not in use.*

*Both the headset and microphone are soundproof. Accordingly, the communication from the aircraft and the words pronounced by the operator cannot be heard by anyone standing next to him.*

*The S-Phone was a duplex transceiver; that is, no switching from* Send *to* Receive *was necessary. Two-way communication took place as with a telephone.*

*The position of the set above the ground and the distance between the operator's body and the aerial were such that ground and body formed 2 right-angled reflectors that reinforced the wave beam in a given direction while at the same time tilting it toward the sky.*

*The set was very directional, and the operator had to face the path of the aircraft. Transmitted signals could not be picked up by any ground monitoring station more than 1 mile away.*

NOTE: *Though the apparatus was miniaturized because it was especially intended for clandestine operations, it had to use the oddity of the huge connector plug in front of the belt buckle because nothing else was available!*

## *Ground-to-Air Radiotelephone Communication: Illustration of the Principle*

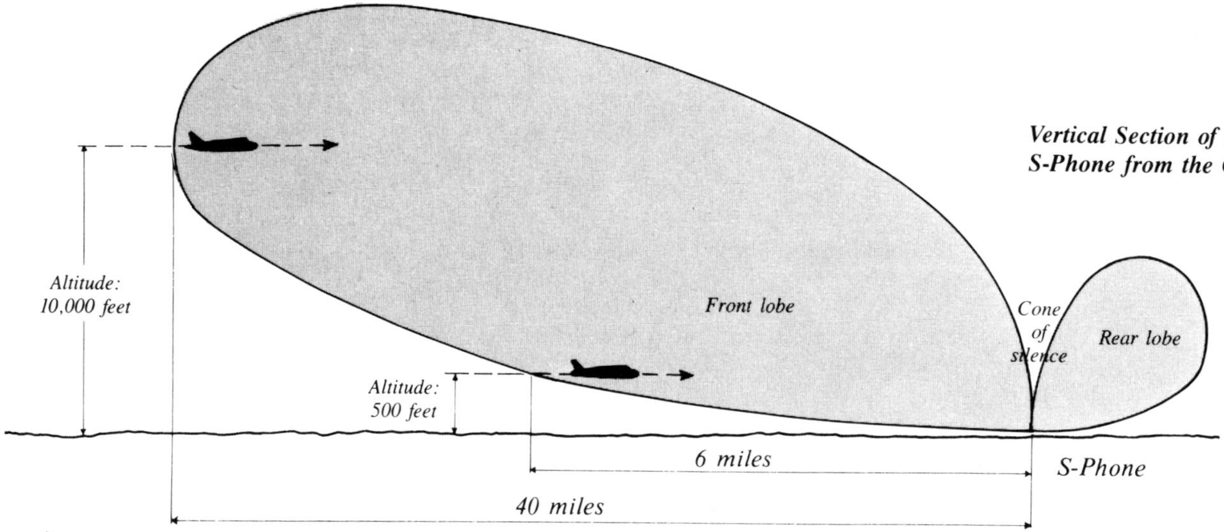


### *Vertical Section of the Beam Transmitted by the S-Phone from the Ground*

*The antenna produces a marked directional effect. The smaller lobe is transmitted behind the operator and the beam is slanted toward the sky.*

*The ground-level range does not exceed a mile. The range achieved at the time of communication with an aircraft is directly proportional to its altitude.*

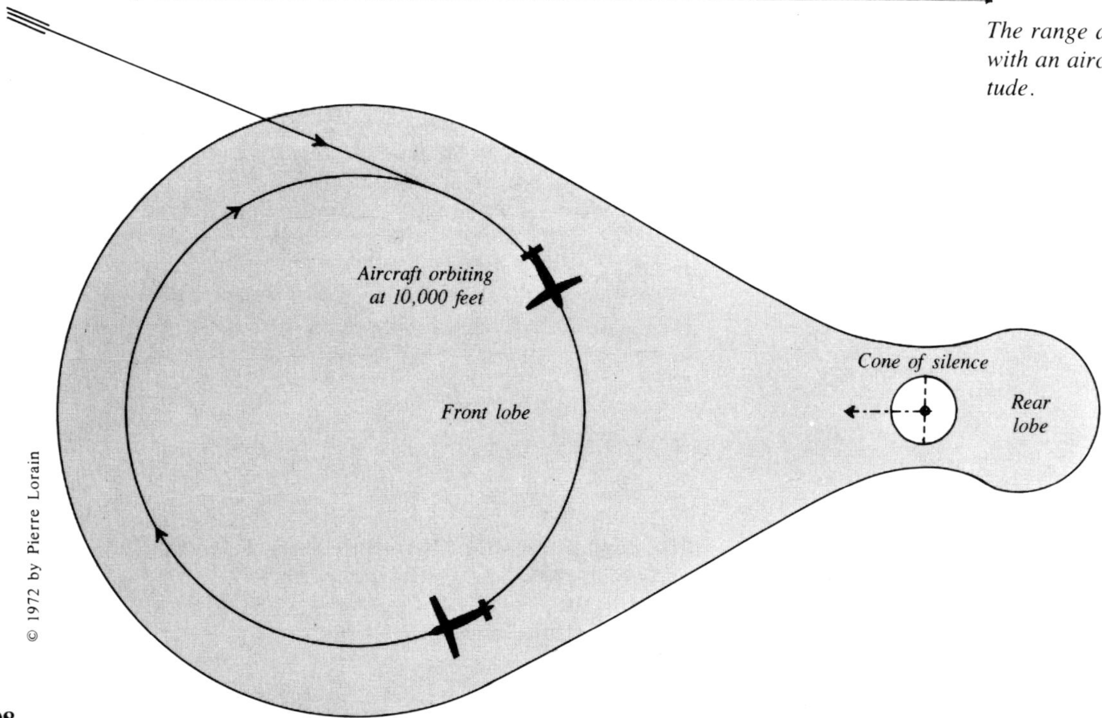


### *Horizontal Section of Beam*

*The cone of silence is above the S-Phone. By having the aircraft turn at a high altitude inside the beam, the agent on the ground can maintain radiotelephone communication as long as necessary without fear of enemy position finding or interception of communications. This procedure was often employed when London wanted to verify the agent's identity. The high quality of the radio communication made it possible to recognize his voice at once.*

**Horizontal Section of Beam Transmitted by the S-Phone from the Ground**

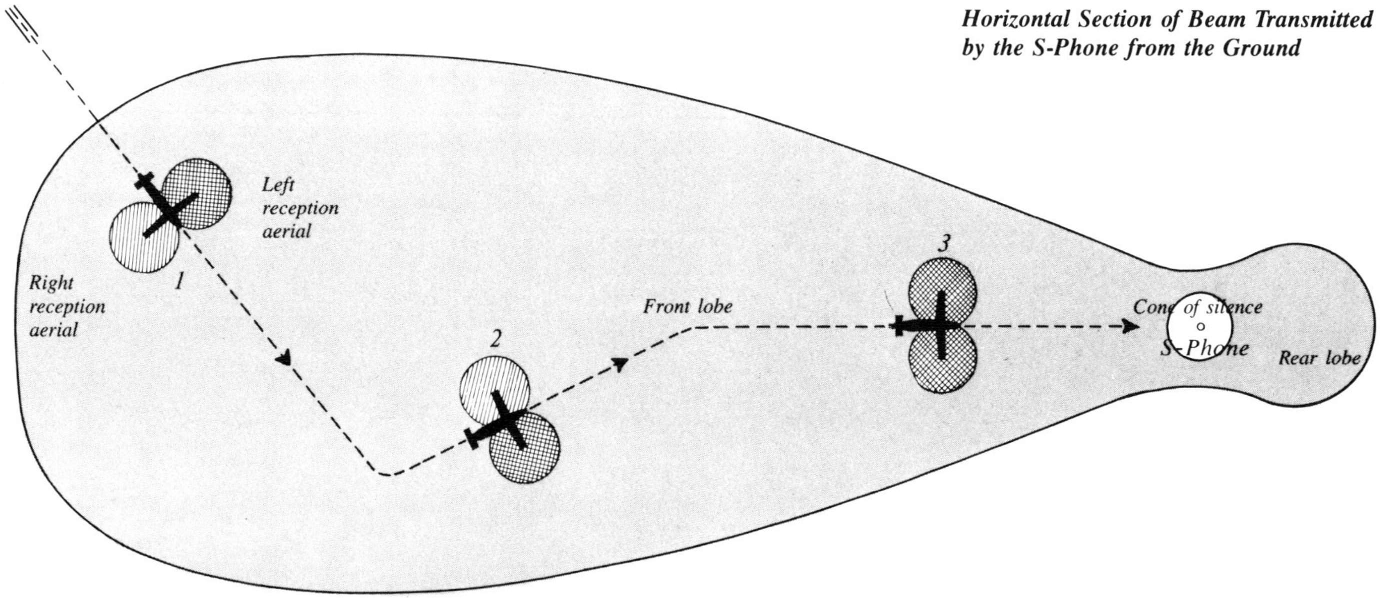




*The range of the S-Phone when used as a homing beacon was almost double the range achieved when used as a radiotelephone. Average range was 12 miles for an aircraft flying at 500 feet (4 minutes at 150 miles per hour) with a maximum of 60 miles for an aircraft at 10,000 feet (24 minutes at 150 miles per hour). Radiotelephone contact could not be established until from 15 to 2 minutes before the plane was overhead.*

*The following indicators show 3 successive positions of the aircraft. The needle indicates the correction of the trajectory needed so the plane can be guided into the axis of the beacon. Distance is not given.*

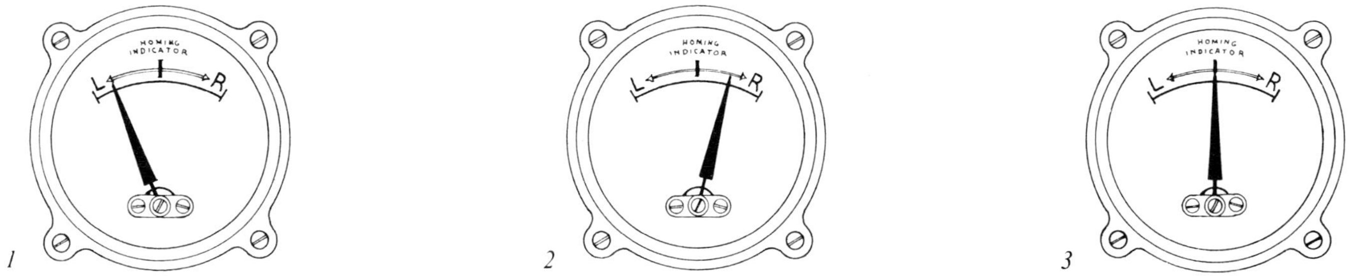


*1. The aircraft is within the beacon's area, but too far to the right. The left aerial is picking up more energy than the right aerial. The pilot has to turn toward the left.*

*2. After the trajectory is corrected, the right aerial now picks up more energy than the left. The pilot must turn toward the right.*

*3. The 2 aerials receive the same amount of energy, and the aircraft is headed straight onto the beacon.*

## "Eurêka-Rebecca" Radar

### *Illustration of the Principle*

NOTE: *The Rebecca on-board interrogator aerial and the Eurêka ground responder aerial are omnidirectional. In addition, the aircraft has 2 very directional receiving aerials which can only pick up a narrow beam directed forward and to the right and left of its route.*

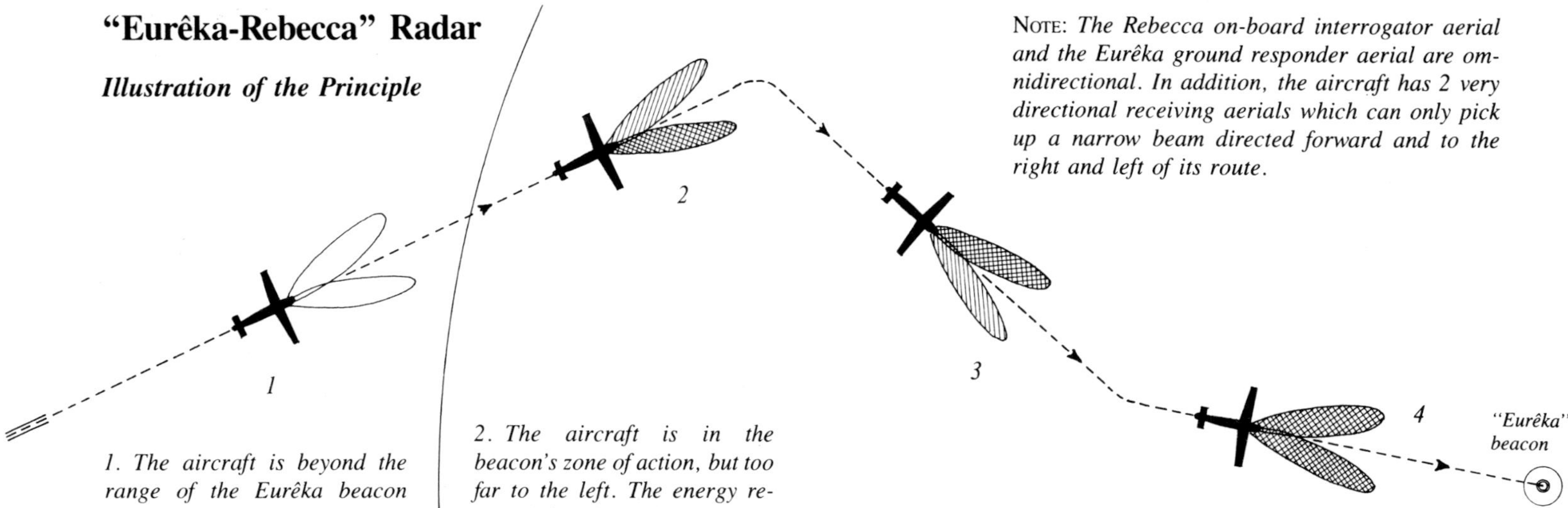


*1. The aircraft is beyond the range of the Eurêka beacon and following a trajectory indicated by the P-10 compass on board.*

*2. The aircraft is in the beacon's zone of action, but too far to the left. The energy received by the right-hand antenna is greater than that received by the left antenna.*

*3. After a correction to the right, the left aerial picks up more energy than the right.*

*4. The aircraft is in the beacon's axis, and the 2 aerials receive equal energy.*

*Rebecca on-board radar indications of 4 successive positions of the aircraft. Two indications are given: the position of the aircraft in relation to the beacon and the distance separating it from the aircraft. The radar screen shown is one third its actual size.*



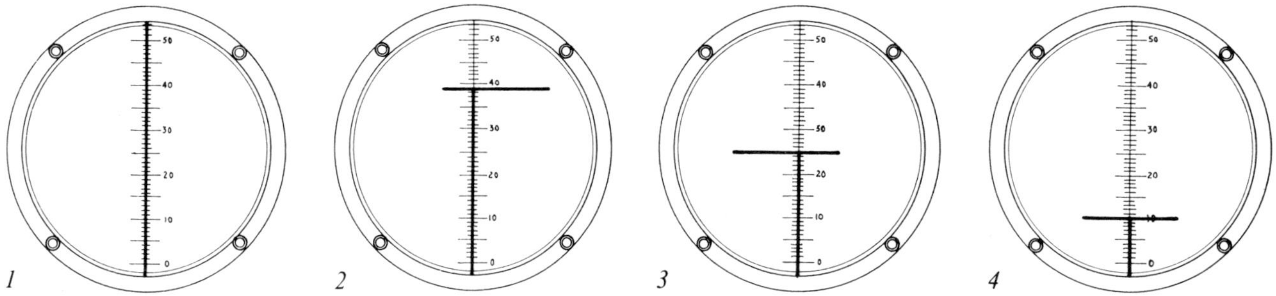


*1. Interrogations without responses. The aircraft is outside of Eurêka's range. The radar screen has a continuous vertical line swept from the bottom to the top.*

*2. Contact. The beacon is located to the right of the aircraft at a distance of 35 miles. The plane must correct its trajectory and head toward the right.*

*3. The line shows that the aircraft is approaching, with its distance now 25 miles, but its trajectory is too far to the left of the beacon.*

*4. The 2 response pulses are equal. The aircraft is advancing straight toward its objective 10 miles away (4 minutes at 150 miles per hour). When the spot hits "0," the plane will be flying directly over the Eurêka beacon.*

### *Measuring Oblique Distance: Illustration of the Principle*

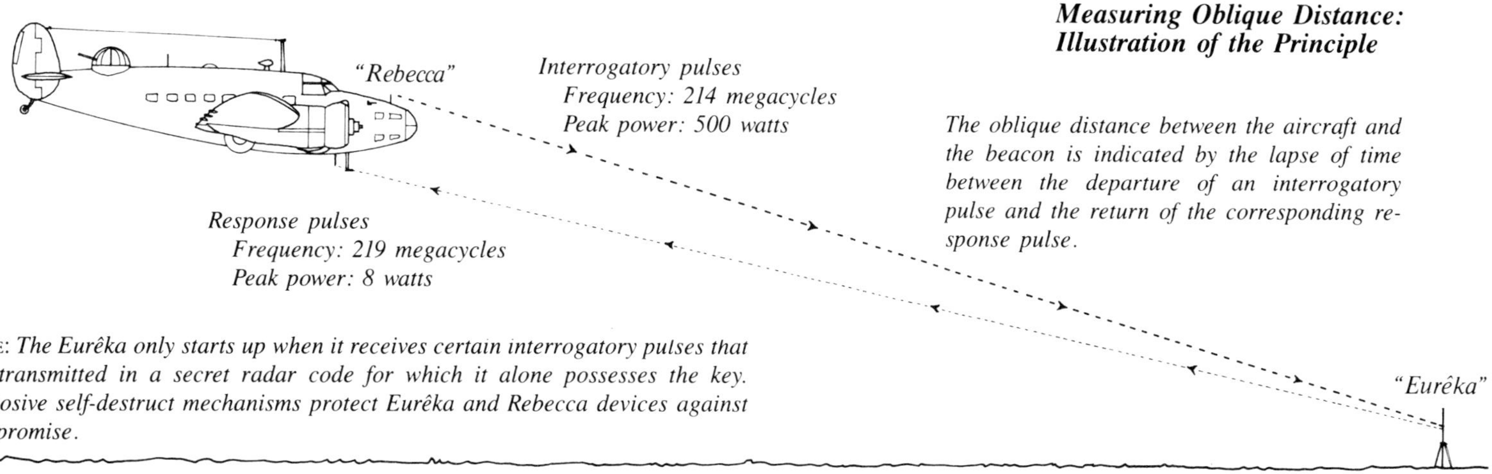


*The oblique distance between the aircraft and the beacon is indicated by the lapse of time between the departure of an interrogatory pulse and the return of the corresponding response pulse.*

NOTE: *The Eurêka only starts up when it receives certain interrogatory pulses that are transmitted in a secret radar code for which it alone possesses the key. Explosive self-destruct mechanisms protect Eurêka and Rebecca devices against compromise.*

## Typical Ground Beacon Layout for Clandestine Parachute Drop Operations

*An open field, 2,000 to 3,000 feet long, is needed. Three red lights (1, 2, and 3) are set up 300 feet apart in the middle of the field, parallel to the direction of the wind. A white light (4), located 150 feet to the side of light 3 and perpendicular to the axis of the parachute drop, completes the marking of the drop zone. This light will flash the Morse recognition letter when the bombers arrive. A Eurêka device is put next to this light. The plane approaches at an altitude of between 500 and 600 feet on the parachute drop axis, against the wind, at the slowest possible speed (100 miles per hour for the Halifax). The bomber drops its containers when it is over light 2.*

*The precision of the Eurêka-Rebecca homing system is remarkable (on the order of less than 750 feet).*

*Colonel Livry-Level, navigator in the 161st Squadron, in his book* Missions dans la R.A.F., *cryptically describes the successful parachuting of 3 I.S. agents using this system in foggy weather.*

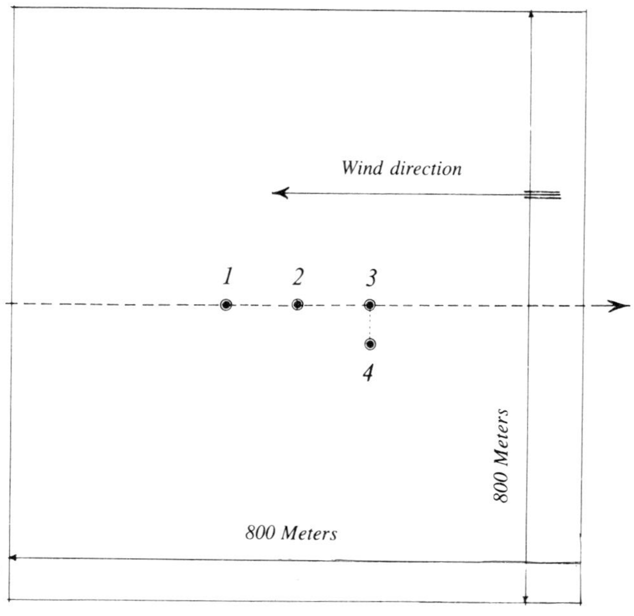


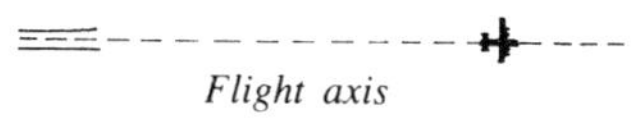


NOTE: *In relatively safe areas, fires made with branches often replaced the three red lamps. Optical signals were made with a car headlight directed toward the sky.*

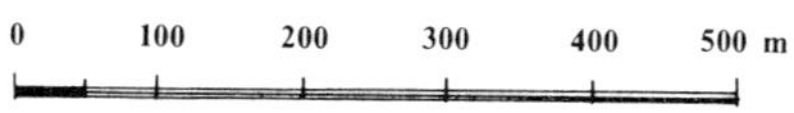

## Compasses

### *High Precision Mark III Prismatic Compass*

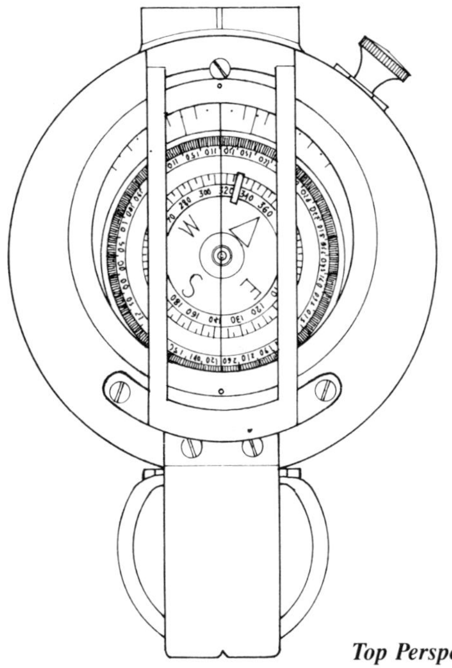


*Actual size*

***Top View***

*The protective cover is closed. The compass rose, marked in degrees, is suspended at the universal joint in oil.*

*These regulation British army compasses were used to carry out precise measurements of landing fields for Lysanders and Hudsons, to mark out beacon axes, and to locate light positions. They could also be used as miniature marine compasses.*

***Top Perspective View***

*The compass is set up. The cover is open, and the optical prism is turned down against the protective glass.*

*When the compass is placed on a stand, readings of an accuracy greater than a degree are obtained.*

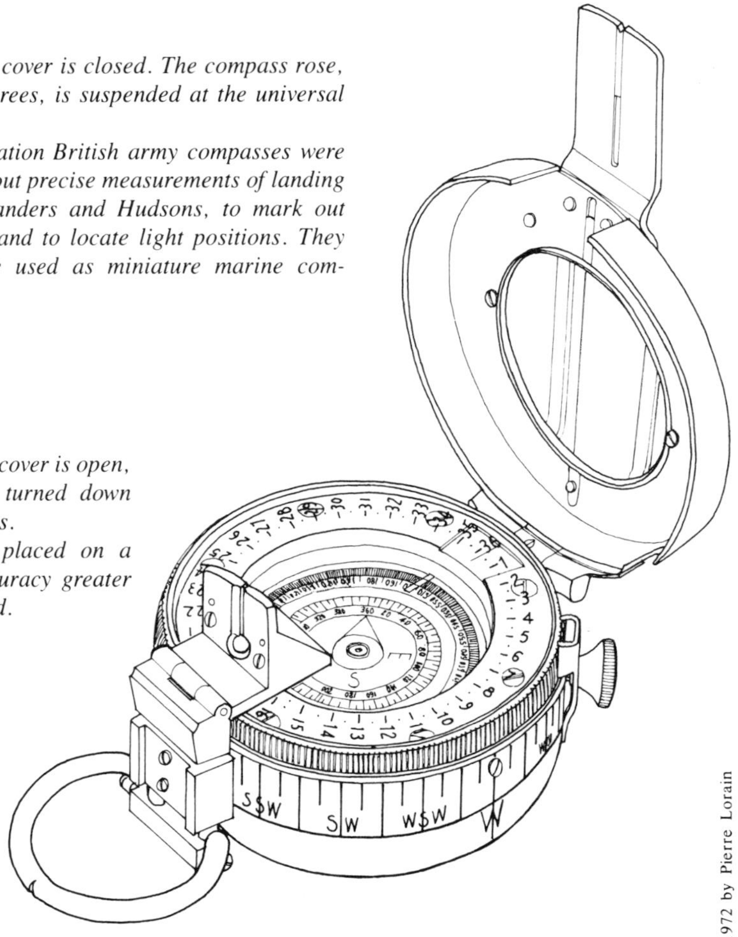


***Top View***

### *Miniature Compass*

*Actual size*

***Side View***

*Weighing a tenth of an ounce, these miniature compasses were concealed inside raincoat buttons. They were a part of all agents' standard survival kits.*

# Chapter VI

# From Order to Delivery: Request, Response, Appearance

After the ground crew members arrived and implanted themselves in the area, their first job consisted of listing fields that could be adapted for parachute drop and pickup operations.

The ideal field for parachute drops was a clear, square field measuring about 2,500 feet on a side, far from any buildings, but with access by road for trucks and carts. Pickup fields for Lysander landings had to have a minimum length of 2,000 feet; for twin-engine Hudsons, 3,000 feet. In either case, the average slope of the field could not exceed 1 percent. The approach areas had, of course, to be free of any large obstructions or high-tension wires.

We will follow the complete procedure of a parachute drop supply mission, from the placement of the order and to delivery. The action takes place in 1944 and is organized by the S.O.E.'s F Section. We will describe the exact methods and techniques employed at the time.

Several weeks after a landing by parachute, the team of 3 agents is installed in the assigned area. Each agent has found safe lodgings known only to him. The team leader has established meeting places and internal communications methods. The radio operator has been able to install 5 transmitting stations spread over a radius of 10 miles from his usual domicile.

Preliminary reconnaissances have spotted 2 fields less than 15 miles away that appear to fulfill the conditions required for parachute drops. Another field further off could serve as a landing field for Lysanders or even Hudsons. These fields are carefully surveyed and recorded on a 1:200,000 Michelin road map.

The team leader then decides to submit these sites for approval from London.

Each field is given a code name as well as a conventional phrase which the B.B.C. will henceforth broadcast during personal messages whenever the S.O.E. headquarters decides upon an operation to that field.

The first field is found on Michelin map 68, fold 18, inside the square where the most important town is Saint-Sébastien. The field's coordinates in millimeters inside this grid square are 47 and 22. Therefore, the field is perfectly defined by the message, "68 18 Saint-Sébastien 47.22."

It is also decided that the field will be called "Delco" and that its B.B.C. message will be, "Mon grand frère est petit" (My big brother is little).

The second field is also found on map 68, in the same fold, but inside the grid square where the town of Dun is located. Its coordinates are 56.28. The field is named "Gicleur" and the corresponding B.B.C. phrase is, "Une âme artiste, c'est moi" (An artistic spirit, that's me).

The third field is called "Carburateur." The corresponding B.B.C. phrase is "La tortue tourmente la tourterelle avec un tournevis" (The tortoise torments the turtledove with a screwdriver). The purpose of such a message would be to pique the enemy's curiosity and perhaps cause him to lose a few hours trying to figure out a phrase in which all the main words start with the same letter and which certainly could have nothing to do with a bird, a reptile, or a tool.

This large field is located in fold 8, inside the grid square where the city of Châteauroux is located. Its coordinates are 02 and 24. Therefore, the radio operator will compose his message as follows, using "DZ" for "drop zone" and "PU" for "pickup" and signing it with his code name:

> PROPOSE DZ SIX EIGHT ONE EIGHT ST SEBASTIEN FOUR SEVEN TWO TWO DELCO BBC MON GRAND FRERE EST PETIT STOP DZ SIX EIGHT ONE EIGHT DUN FIVE SIX TWO EIGHT GICLEUR BBC UNE AME ARTISTE C EST MOI STOP PU SIX EIGHT OH EIGHT CHATEAUROUX OH TWO TWO FOUR CARBURATEUR BBC LA TORTUE TOURMENTE LA TOURTERELLE AVEC UN TOURNEVIS STOP XXX TWEEDLEDUM.

The message contains 257 letters to which must be added 3 dummy letters (XXX) to have, along with the signature, a text of 54 groups. The radio operator must absolutely add his "true check" to confirm his identity.

Let's assume he has chosen the word "Tweedledum" and that, on the other hand, "Jabberwock" would mean he is working under enemy control. He must also place the first key group drawn from the one-time pad at the beginning of the message, after changing one of its 5 letters, and repeat this group with the error at the end of the message. The complete message will thus comprise 56 groups.

The silk handkerchief tableau and microfilmed sheet 1, marked "Outstation to Home Station," are taken from their respective hiding places. The first key group from the one-time pad is BDSKS. Because it has been arranged to produce an "error" with one of the letters if the station has not fallen into enemy hands, BDRKS will be used. The next 54 key groups are then written out on the rough draft with the plaintext underneath. The tableau provides the cipher letter as determined by the key and the plaintext letter:

trunk with 3 grenades and a Colt pistol. The quartz crystals in their canvas pouch are concealed under some old rags.

About 10 minutes before transmission time, the radio operator arrives, pushes his bicycle into the shed, and extracts the message from the pump. Once the door of the attic is blocked with steel angle irons, he takes out the suitcase and the crystals.

He methodically plugs in the key and the headset, and connects the aerial to the instrument's terminal. The operations program shows a frequency of 5,570 kilocycles for this first transmission. The proper crystal is placed in its holder. The 230-volt supply line is attached to the set, and the lamp-socket adapter is connected to the old lamp hanging from the rafters. The house current power switch is turned to *On*.

While the set is warming up, the radio operator generally loads a bullet into the Colt even though he runs no

| | | | | | | |
|---|---|---|---|---|---|---|
| B D R K S | T V N S S | A P Z W K | U A P S T | K M S A F | . . . . . | B D R K S |
| | *p r o p o* | *s e d z s* | *i x e i g* | *h t o n e* | . . . . . | |
| | Z F P N Y | L X D V A | Z Z X F H | O M Y U O | . . . . . | |

The final draft of the message is then written on another sheet: GR 56 = BDRKS ZFPNY LXDVA ZZXFH OMYUO . . . BDRKS AR. (The AR is a standard radio sign meaning end of message.)

The radio operator cuts the 55 key groups out of sheet 1 and burns them along with the rough draft of the encipherment. The ashes are crushed and scattered about. The tableau and one-time pad are returned to their hiding places. The enciphered message is rolled up and concealed in—for example—the barrel of a bicycle pump. It will be sent at the next departure, perhaps the following day at 3:15 P.M., to station 1, some 5 miles away.

This station was carefully installed during the previous days in an attic. The aerial hangs between 2 pieces of lumber, camouflaged as a clothes line. The suitcase transceiver, an A Mark III, has been placed in an old

risk of being spotted by the Germans during a first transmission.

About 15 seconds later, the characteristic "waterfall" sound is heard in the earphones: the set is working. The switch is turned to the red "transmission" dot and the key is depressed. The milliammeter needle points to 50 milliamperes. The anode tuning knob is quickly turned to the left. The needle suddenly falls downward, pointing to between 10 and 15 milliamperes. Without taking his finger off the key, the operator moves the aerial switch successively to positions *2*, *3*, and *4*. The milliammeter needle moves back up and reaches the red mark. Then, for the final tuning, the operator turns the anode tuning knob a bit toward the left until the needle moves very slightly downward. The set is now tuned, and the key is released at once. The tunings take about 10 seconds and do not have to be readjusted for future transmissions on the same frequency and with the same aerial.

Two minutes before the transmission, at 3:15 P.M., the radio operator limbers up his fingers and places the telegram text where he can read it, his left hand enclosing the miniature cap of the key.

At 3:15, his right hand begins to tap out: BGT QTC 1 GR 56 = BDRKS ZFPNY LXDVA . . . BDRKS AR.

The transmission lasts 2 minutes and 15 seconds. Five minutes later, the set is folded up and the suitcase, pistol, grenades, and quartz crystals are put back into their hiding places. The telegram has just been burned in an ashtray.

The radio operator gets back onto the road, carrying some eggs and a salad in a basket to justify his trip, if that should become necessary.

At the secret home station 53-B, some 50 AR-88 receivers wait for messages. At 3:15, an operator starts the recording device. The signals arrive strong and clear. She quickly writes them down with her pencil. The telegram is immediately taken to the code room where it goes through the regular channels: deciphering and teletype transmission to Baker Street. The message is preceded by the date and time of reception, as well as the phrase "true check present." (If the agent had indicated that he was operating under duress, the phrase "bluff check present" would precede the message.) A quick comparison of the transmission with the sample recording verifies the operator's identity. The network is operating normally.

The coordinates for the proposed airfields will be transmitted to the air liaison office so that aerial photographs can be obtained.

That same evening at 10 P.M. the radio operator locks the door of his shabby hotel room, stretches the aerial out across the room and takes a miniaturized Biscuit receiver out of his suitcase. The set is quickly connected to the battery. The headset and aerial wire are plugged in and coil block 2 is put in place. The conversion chart shows that the home station's frequency is approximately 120 on the dial. This number is brought over against the reticule, the regeneration control is turned up to the limit of the whistling, and then the volume is turned down as low as possible so that the Morse signals cannot be heard outside the room.

At the exact time specified, tuning signals from a powerful transmitter at the home station are heard: VVV DE UZM QTC = RDQ 3 EWT 2 OXG 2 UGH 1 AS.

Since the network's call letters for that day are UGH, the radio operator must wait until the home station automatically passes on the 7 messages meant for the first 3 networks. Each message is repeated twice at normal speed and then faster a third time so that the recipient can read it over. At last there arrives: UGH QTC 1 GR 8 = BWOPS KNPAC DDSKP TZZXK NPTRA TLPAN GSITI AESYU AR.

The radio operator cuts out the first 8 groups from the key sheet marked "Home Station to Outstation" and writes them over the ciphered groups of the telegram. In a few minutes, using the tableau, the operator deciphers the letters into the following message: *Your five six acknowledged Stop Good luck Stop X.*

The operator uses his last cigarette to burn up the telegram and the 8 key groups. He takes down the aerial and puts the receiver back into the suitcase. He is no longer alone: henceforth, a tenuous but safe thread connects him with far-off England.

A few days later, a message from London declares that the 3 fields are officially approved. It is now time to place the orders: the deliveries will take place during the coming lunar periods and will function on a strict priority plan.

It is decided to request for the next full moon a first delivery of 6 containers of explosives, sabotage material, and submachine guns. Since the "H"-type containers are standardized, it is decided that 2 loads of Sten submachine guns, 2 of explosives, and 2 of sabotage materials will be delivered to the Delco field.

The radio operator prepares his telegram as follows, using the standard numerical designations for "H"-container material: *Send two H one Stop two H two Stop two H five Delco Stop Tweedledum.*

At the beginning of the lunar period, another telegram from the home station announces that the order will be dropped the following night: *Stand by for Delco drop Stop Flash Love* [letter L in phonetic alphabet] *Stop.*

The next day, after the 7:15 P.M. B.B.C. French language news broadcast, listeners in several outlying towns strain their ears to try to pick "personal messages" out of German jamming signals. Someone manages to make out "Mon grand frère est petit." By 9:25, the B.B.C. is no longer jammed and everyone clearly hears the message, "Mon grand frère est petit." The parachute drop is confirmed for the night.

The teams meet with their mission leader and his assistant. At 11 P.M. an old gasogene truck jiggles toward the field. At the same time, groups form in neighboring towns and head for the field. Around midnight, some 30 men are present. Eight armed men take up positions to preclude unwanted access to the field. Eighteen men, formed into 3 teams of 6 for carrying away the containers, gather along the line of brush bordering the field. The mission leader and his assistant locate the sites for the signal and beacon lights. Without a sound, each man huddles along the edge of the field; the wait begins under the pale light of the moon.

The mission leader is wearing a raincoat which conceals the S-Phone and the batteries hooked onto his webbed belt. About 12:30 A.M., he takes the headset and microphone out of their canvas sack, unfolds the small, funny-looking aerial, and attaches it to the set. Alone, he heads toward the middle of the field where he starts up the vibrator and faces north. A loud "waterfall" sound fills the earphones. The wait continues far into the night. It is almost 1 A.M. From time to time, the mission leader looks at his compass and alters his position slightly.

At this very moment, a completely black Halifax is flying over French soil at an altitude of less than 300 feet. Its compass shows a magnetic heading of 150 degrees as it slips above the ground at close to 250 miles an hour, surrounded by the din of 5,800 horsepower.

The plane has just crossed the Loire and has bent its course toward the south somewhat. At 11:15 P.M., it left the secret Tempsford airfield with just the following orders: "Drop your 6 containers on the field for which you have the coordinates and aerial photographs. The reception committee will be waiting there between 1 and 2 in the morning and will signal the letter *L* in Morse code with a white light. The head of the reception committee will have an S-Phone to guide you and help you find the field's exact location. He will contact you by radiotelephone using the code name 'Paddy.'"

It is 12:50 A.M. According to the navigator's calculations, the aircraft should fly over the field at 1:10 A.M. Over the intercom, the navigator instructs the pilot to climb to 5,000 feet while maintaining the heading at the same time. The S-Phone's visual indicators have still not budged. The pilot rises another 1,000 feet. The navigator scrutinizes the receiver dial. At 12:55, the needle moves slightly to the left: "There they are!" The heading is maintained, and as the seconds go by, the needle bends more and more to the left. The pilot reduces his speed to 150 miles an hour and corrects his course 10 degrees to the left. The needle reacts immediately by going beyond its middle position. Another correction stabilizes the needle in the vertical position. It is now time to redescend. The direction indicator moves again at the altitude of 1,000 feet. The navigator switches on his transmitter and calls, "Hello Paddy, hello Paddy. How are you receiving me? Over." There is no response from the ground as yet, but the needle now moves at the slightest action by the pilot on the plane's controls. The plane is speeding straight toward its objective.

On the ground, the mission leader has just listened to the first call from the plane. One or 2 minutes afterward, 2-way communication can be established. The leader raises his arm and the men in charge of the beacon system and signaling take their respective positions in the middle of the field even though no sound of engines can yet be heard.

On board the aircraft, the chronometer reads 1:05 A.M. Until now, the needle of the signals level indicator had stayed on position 12, but suddenly drops to 6 when the still weak and far-off voice from the ground says, "Hello Paddy, hello Paddy, receiving you loud and clear. Can you hear me? Over."

"Roger, Paddy, 5 by 5." Two-way communication is established.

The pilot informs the ground that he wants to make a preliminary pass over the field and then drop the containers in 2 passes. At the same time, the men on the ground hear the sound of engines; it grows rapidly. The 3 red lamps light up, and the white light, pointed toward the sky, begins to flash: dot, dash, dot, dot.

On board the aircraft, the navigator has asked the pilot to reduce the sensitivity of the position indicator. A slight correction of the trajectory and then, "Okay, Paddy, I see your lights. I'm going to pass over the field within 1 minute."

The altitude of the aircraft is now 500 feet, and its speed has been cut to 120 miles an hour. With flaps lowered, the plane flies straight onto the axis of the ground lights. It is traveling against the wind.

The ground lights file under the navigator's eyes. The S-Phone indicator needle jumps roughly from left to right as considerable interference muffles the mission leader's voice.

On the ground, the aircraft's shadow has just crossed over the field, accompanied by the roar of engines. The enormous plane continues to speed away, turns, disappears but comes very quickly right back over the field. The navigator announces the first drop. The roar of the engines swells enormously and 3 white flowers fill the sky. Five seconds later, 3 heavy thuds are heard and the black cylinders lay stretched out toward the end of the field.

The plane passes over again, causing everyone to stoop down even more. Three more cylinders fall a little short, close to the brush. The men fall upon them. The navigator announces a final pass to drop a personal package. The aircraft flies over the field for the last time, turns, and heads north. The navigator has hooked up the S-Phone to the intercom, and all the crew members, from the pilot down to the tail gunner, shout words of encouragement down to the ground team.

The mission leader continues to exchange some pleasantries with his fellow countrymen, but soon the friendly voices weaken and vanish. The marvelous visit has come to an end. The men, still overwhelmed, mechanically and silently undertake the maneuvers taught them for the removal of containers.

Any trace of the operation would have disappeared by dawn; 1,500 pounds of arms and explosives have safely arrived on French soil.

To put the enemy on the wrong track regarding its mission, the Halifax drops several thousand propaganda leaflets over the nearest town and hedgehops back to its secret haven.

# Chapter VII

# Manna from Heaven

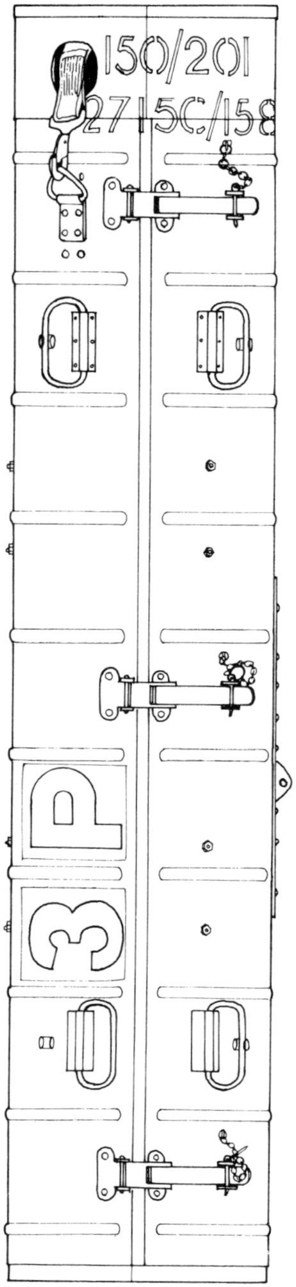


# CONTAINERS

## Type C

*The Type C container is a sheet-metal cylinder reinforced by ribs. It opens in 2 along its length and is closed with 3 latches.*

*At left:*

*Container showing the latches and the line along which it opens.*

*The parachute housing is on the upper part of the container. A webbed strap with a clip links the parachute to the body of the container. Four men carry the container by passing boughs through the handles.*

*C containers could be divided into 3 equal compartments with wooden disks, according to the size of the equipment to be housed there.*

*The drop identification numbers, here a 3 and a* P, *are printed in black on white paper which was stuck onto the container.*

*At right:*

*This view of a C container shows the side it was hung from when inside the plane.*

NOTE: *C containers had a frame about the size of a human body. Once emptied, they made excellent targets for shooting practice. Orders called for the immediate destruction of these compromising parcels.*

0 10 20 30 40 50 60 cm

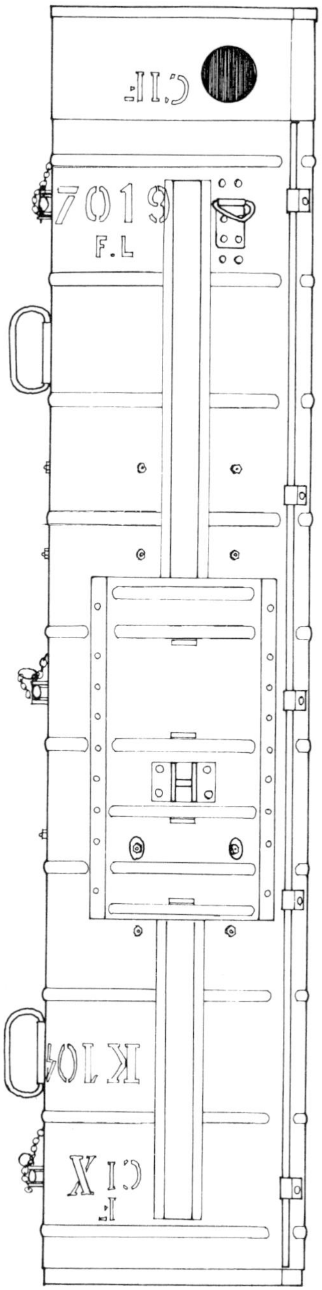

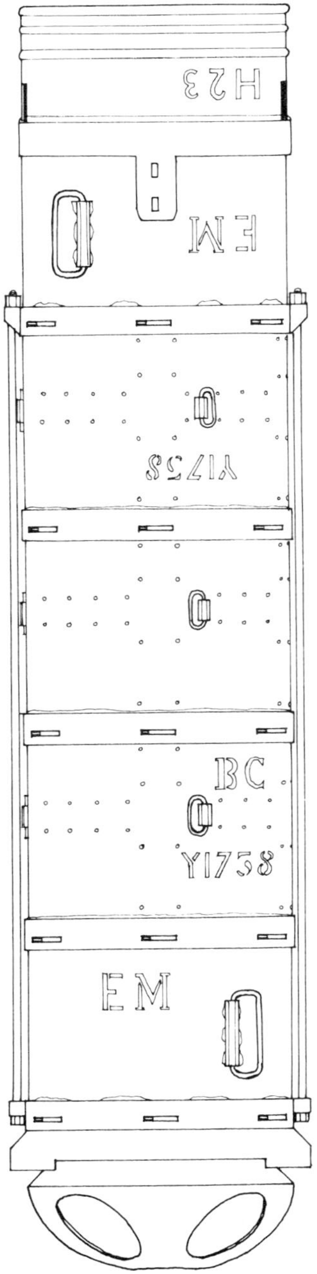


# Type H

*This container is made up of 5 sheet-metal drums piled one on top of the other and held together by 2 steel rods diametrically opposite one another and screwed on at the ends.*

*At left:*

*The H container set up. There are, from top to bottom, the parachute housing, the 5 drums, and finally, a hollow spherical sheet-metal dome that serves as a shock absorber. This system was reasonably efficient, but the H container was certainly a lot less sturdy than type C. If the parachute failed to open, the H container could break open when it hit the ground. (The delicate tyre bursters were only packaged in type C containers.)*

*At right:*

*This is a side view of a single drum and a top view of the drum cover with its ingenious latch system. The small knotted cord was used to pull the cover off. Through the small rings along the length of the drum were passed webbed straps for carrying the drum.*

*Type H containers could carry 5 different standard loads:*
*H 1: Explosives and accessories*
*H 2: Sten submachine guns*
*H 3: Weapons*
*H 4: Incendiary material*
*H 5: Sabotage material*

*A very short message was all that was needed to place an order: "Send 3 H 2 and 6 H 5."*

*All H containers were painted dull black, with white stencil letters.*

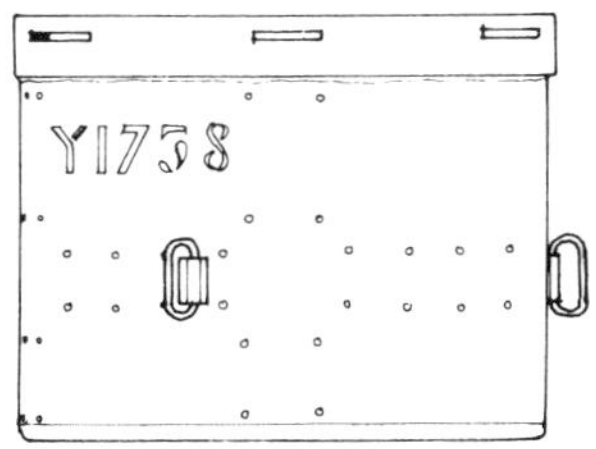


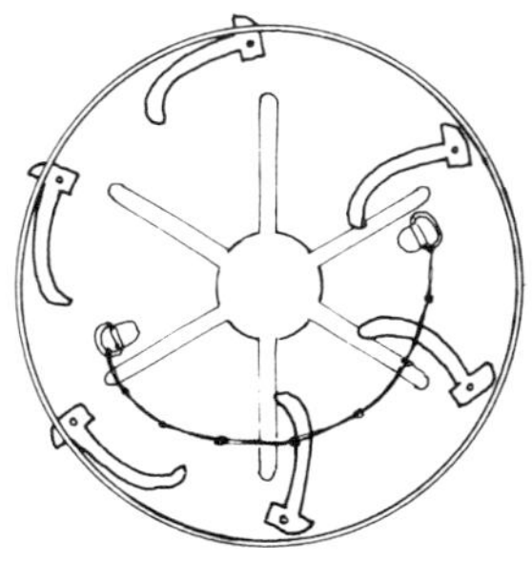

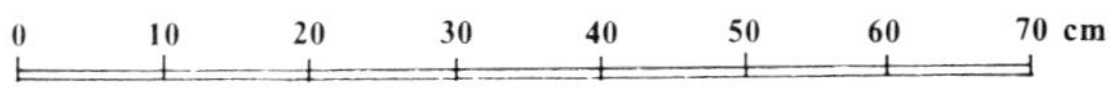

# PISTOLS AND REVOLVERS

## American Colt Automatic Pistol Model 1911-A1

***Caliber: .45***

*Scale: 1/2*

STANDARD FEATURES

*Overall length: 8.62 inches*
*Barrel length: 5 inches*
*Weight unloaded: 2 pounds, 7.5 ounces*
*Muzzle velocity: 850 feet per second*
*Muzzle energy: 380 foot-pounds*
*Breech is locked until bullet has left barrel*
*Operation: short recoil*
*Finish: dull blueing*
*Grips: wood*

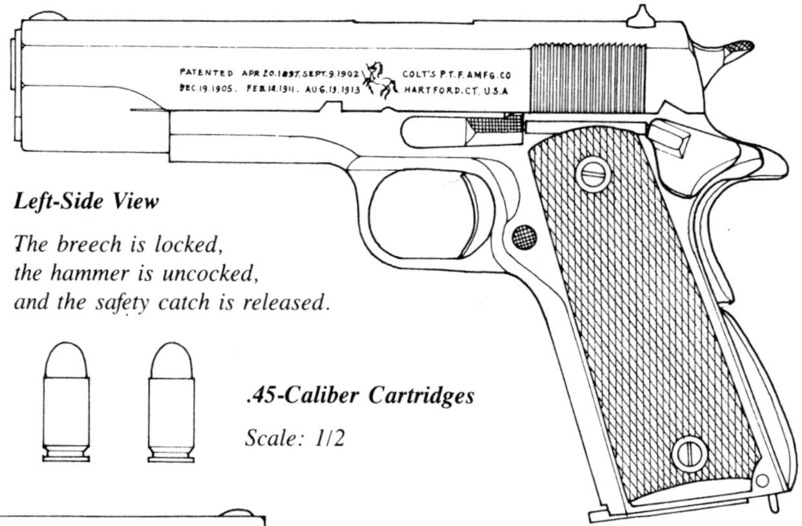


***Left-Side View***

*The breech is locked,*
*the hammer is uncocked,*
*and the safety catch is released.*

***.45-Caliber Cartridges***

*Scale: 1/2*

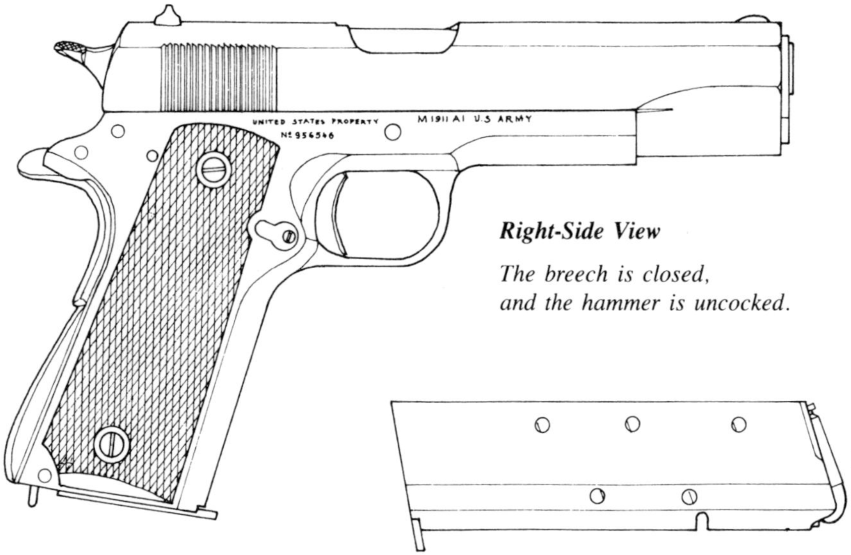


***Right-Side View***

*The breech is closed,*
*and the hammer is uncocked.*

NOTE: *Without a doubt, the Colt is the most famous of pistols. A regulation weapon in the French army during both World Wars, it remained in service until about 1960. It has always been a regulation arm for the United States army. Great Britain used it to a great extent between 1940 and 1945, and many were dropped by parachute. The gun is very sturdy, and its action is completely safe. Disadvantages include a magazine of too low a capacity and its firing of a very large-caliber cartridge that is heavy and outmoded compared to the 9-millimeter cartridge (approximately 0.38 caliber).*

***Seven-Round Box Magazine***

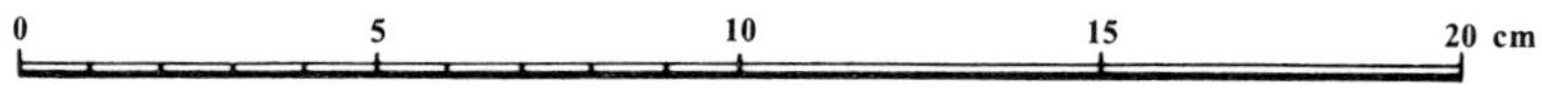

# Canadian Browning Automatic Pistol No. 2 Mark I

***Caliber: 9-Millimeter Parabellum***

*Scale: 1/2*

Standard Features

*Overall length: 7.75 inches*
*Barrel length: 4.65 inches*
*Weight unloaded: 2 pounds, 3 ounces*
*Muzzle velocity: 1040 to 1500 feet per second, depending on cartridge employed*
*Muzzle energy: 310–400 foot-pounds*
*Bolt closed and locked until bullet leaves the barrel*
*Operation: short recoil*
*Finish: completely smooth, dull black*
*Grips: wood*

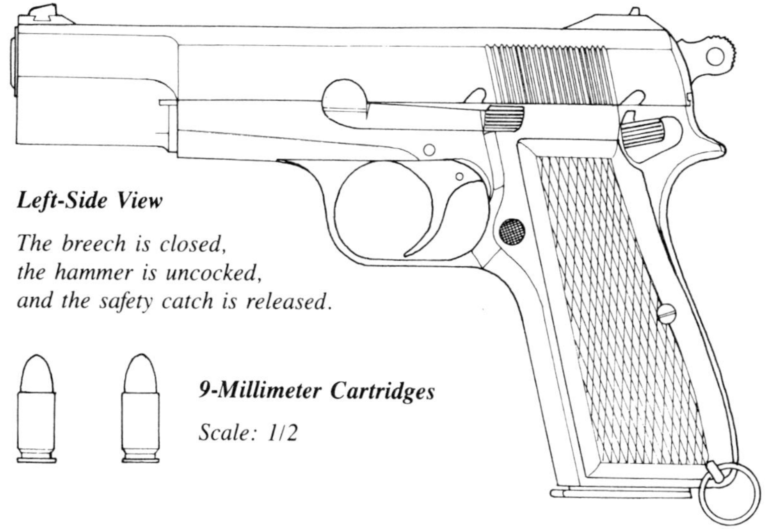

***Left-Side View***

*The breech is closed, the hammer is uncocked, and the safety catch is released.*

***9-Millimeter Cartridges***

*Scale: 1/2*



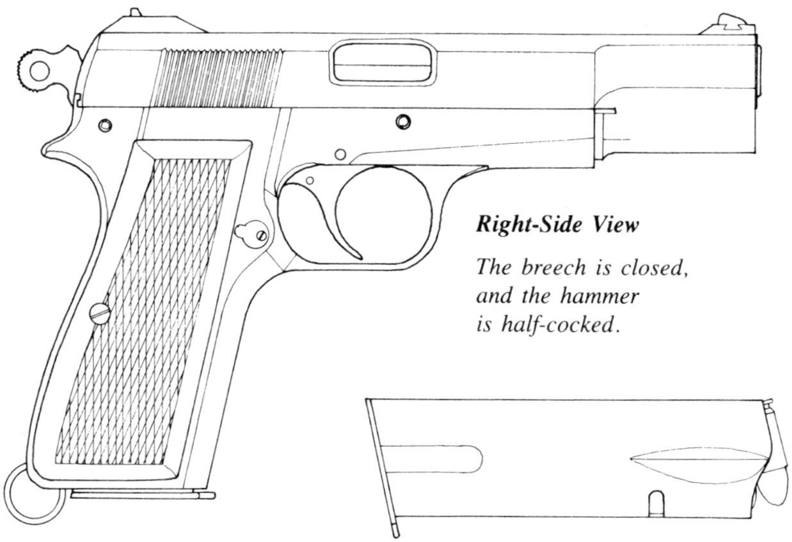

***Right-Side View***

*The breech is closed, and the hammer is half-cocked.*

Note: *This pistol of Belgian origin was designed in 1935 and manufactured under license in Canada by the John Inglis Company of Toronto during the war. In the hands of S.O.E. and S.A.S. Commando experts, it was a formidable weapon.*

*Its sturdiness enabled it to fire the heaviest submachine gun ammunition. Its magazine capacity was unusual.*

***13-Round Box Magazine***

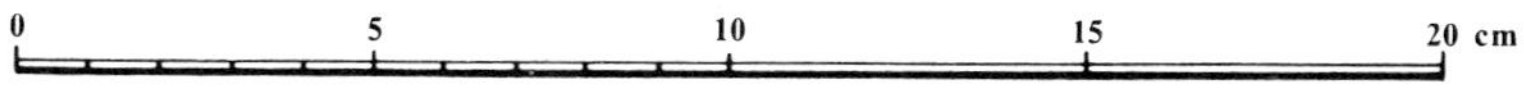

# British Enfield Revolver No. 2 Mark I

***Caliber: British Service .380***

*Scale: 1/2*

Standard Features

*Overall length: 10.25 inches*
*Barrel length: 5 inches*
*Weight: 1 pound, 11 ounces*
*Muzzle velocity: 625 feet per second (English cartridge)*
*Muzzle energy: 110 foot-pounds*
*Action: single or double*
*Capacity: six .380 cartridges*



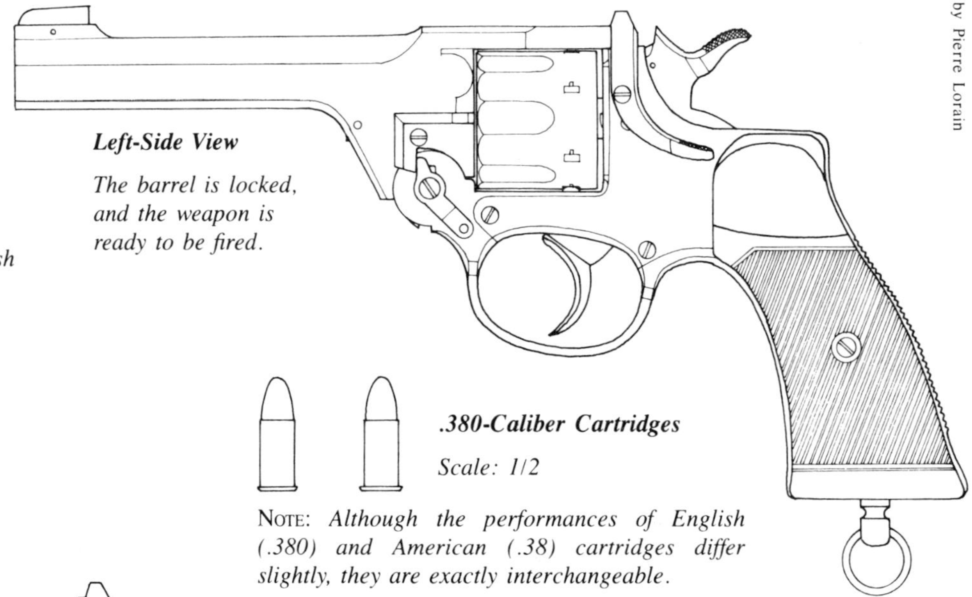

***Left-Side View***

*The barrel is locked, and the weapon is ready to be fired.*

***.380-Caliber Cartridges***

*Scale: 1/2*

Note: *Although the performances of English (.380) and American (.38) cartridges differ slightly, they are exactly interchangeable.*

*Black-blued finish. Black plastic grips with thumb recess.*

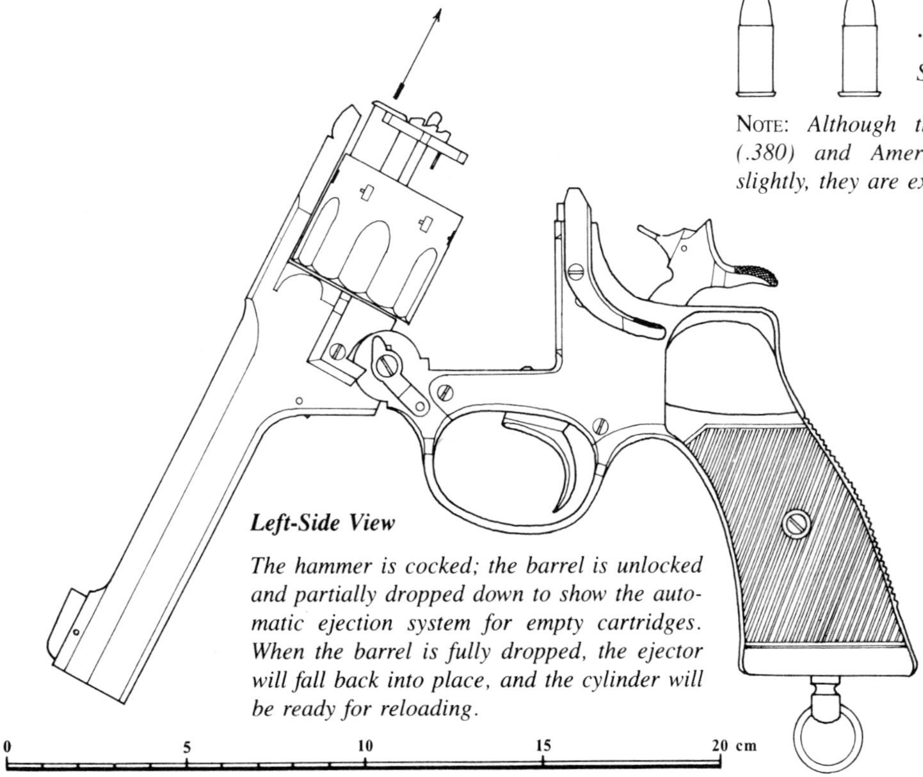


***Left-Side View***

*The hammer is cocked; the barrel is unlocked and partially dropped down to show the automatic ejection system for empty cartridges. When the barrel is fully dropped, the ejector will fall back into place, and the cylinder will be ready for reloading.*

Note: *Another model, introduced in 1938 and intended for the Tank Corps, is identical except that the spur of the hammer has been eliminated and the single-action cocking notch has been smoothed over. Therefore, it could only be fired in double action. Worn under the arm in a holster of supple leather, there was no danger of its hooking into something when the agent pulled it out. The accuracy of double-action firing is negligible beyond 15 or 20 feet. However, for the agent's immediate self-defense, this model offered an unquestionable advantage. This revolver, sometimes called the Enfield Commando, was marked* No. 2 Mark I* *and weighed about an ounce less than the No. 2 Mark I.*

# British Webley Mark IV Revolver

## *Caliber: British Service .380*

*Scale: 1/2*

STANDARD FEATURES

*The weapon approximates the Enfield in size and weight and has the same general features. None of these weapons' parts are interchangeable, but they use the same cartridge.*

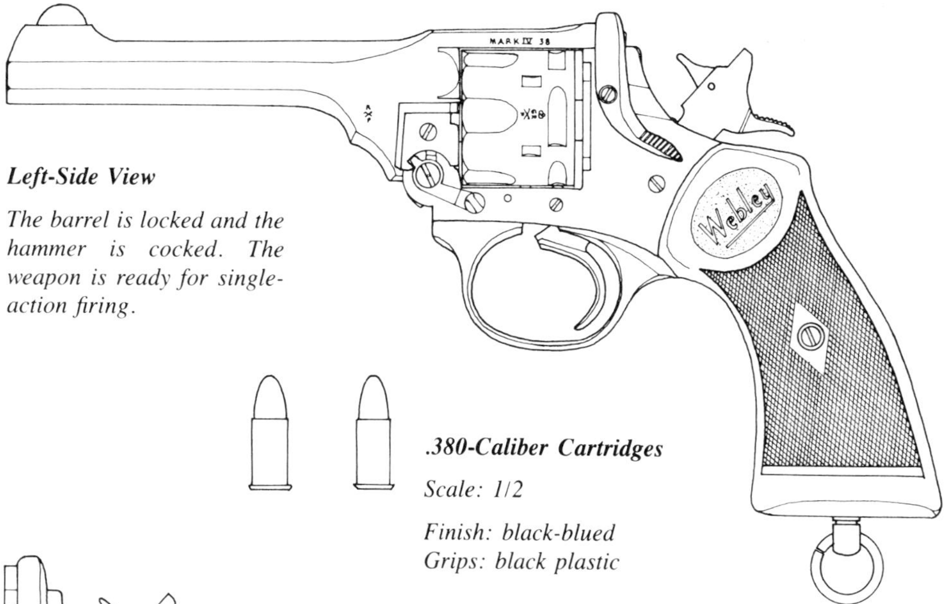


***Left-Side View***

*The barrel is locked and the hammer is cocked. The weapon is ready for single-action firing.*

***.380-Caliber Cartridges***

*Scale: 1/2*

*Finish: black-blued*
*Grips: black plastic*



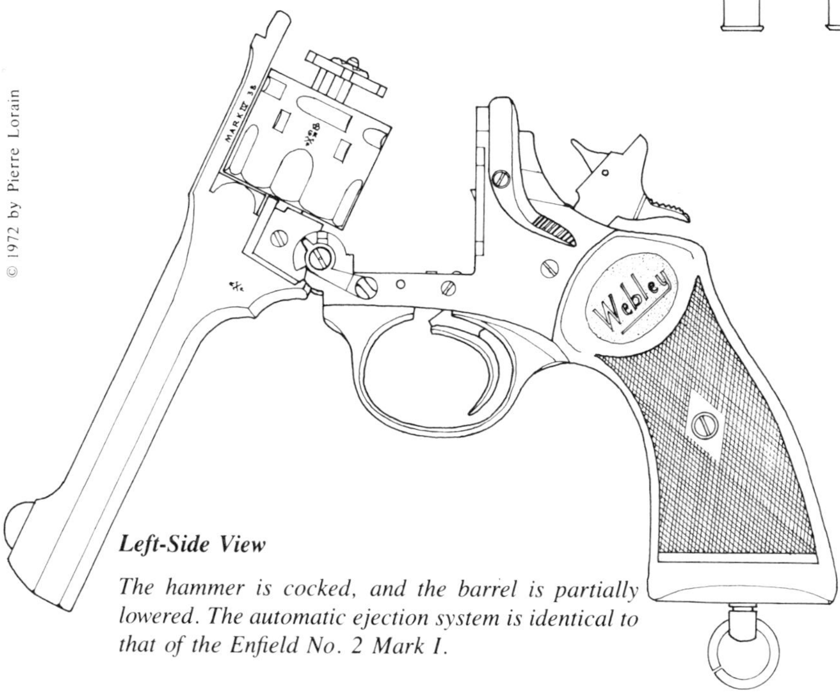


***Left-Side View***

*The hammer is cocked, and the barrel is partially lowered. The automatic ejection system is identical to that of the Enfield No. 2 Mark I.*

NOTE: *This revolver was regulation in the British army. It was widely distributed in French Resistance forces. It was a very well-made gun, similar to the Enfield, though it did not fit the hand as well because of the shape of the stock butt. In spite of the interest in designing a revolver for special operations, these models were considered outmoded on the Continent. Requests for automatic pistols to replace them became more and more insistent.*

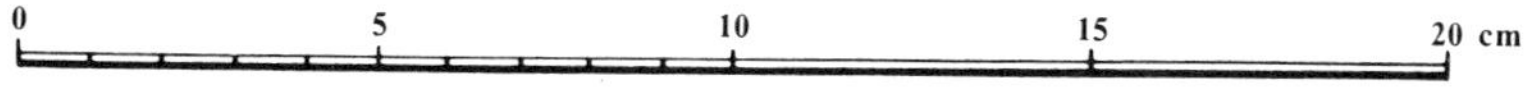

# American Smith & Wesson Revolver

***Caliber: British Service .380***

*Scale: 1/2*

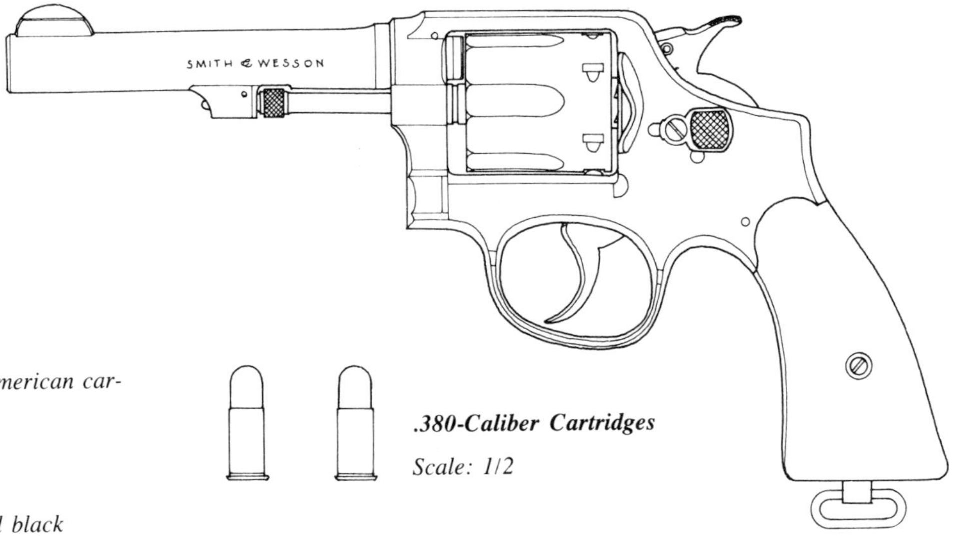




STANDARD FEATURES

*Overall length: 10.2 inches*
*Barrel length: 5 inches*
*Weight: 1 pound, 13 ounces*
*Muzzle velocity: 775 feet per second (American cartridge)*
*Muzzle energy: 110 foot-pounds*
*Action: single or double*
*Capacity: six .380 cartridges*
*Finish: completely smooth, painted dull black*
*Grips: smooth wood*

***.380-Caliber Cartridges***

*Scale: 1/2*

***Left-Side View***

*The cylinder is locked into place, the hammer is uncocked, and the weapon is ready for firing.*

NOTE: *Along with the Colt automatic and the Webley Mark IV revolver, the Smith & Wesson revolver was one of the weapons most widely distributed to Resistance forces. Although this revolver was American in design and manufacture, it fired the regulation British service .380 cartridge and was widely used in the British army, like the Enfield and Webley revolvers.*

*This weapon is excellently built. However, like all the revolvers described here, it possesses the disadvantage that it fires a special cartridge, different from those used in submachine guns.*

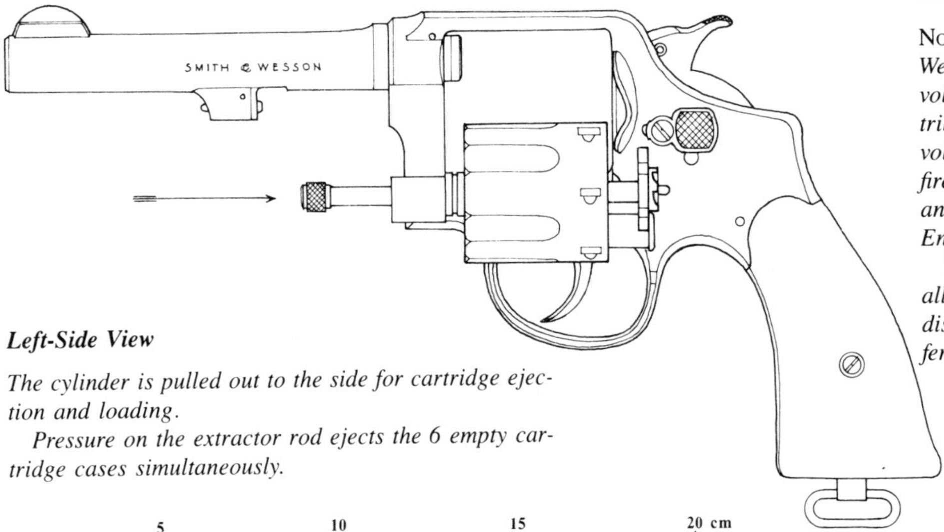


***Left-Side View***

*The cylinder is pulled out to the side for cartridge ejection and loading.*

*Pressure on the extractor rod ejects the 6 empty cartridge cases simultaneously.*

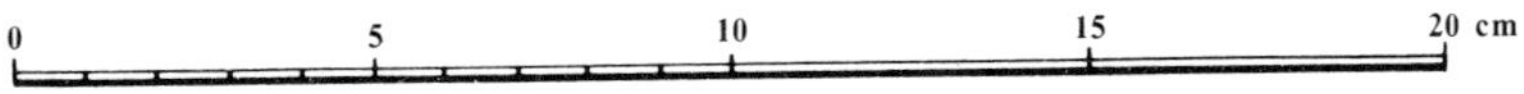

Colonel Maurice J. Buckmaster, in his book *Specially Employed* (London, 1952), points out that of the 418,083 weapons delivered to France during the war, 57,849 were pistols and revolvers. This represents one-seventh of the total, which would be an unusually high proportion for conventional war operations, but is logical for subversive operations because it is easy to conceal such weapons.

However, the S.O.E. could not keep up with the demand for automatics. Everyone wanted a Colt and accepted a revolver only grudgingly. And wrongly, it seems. But that is another story.

Starting in 1944, when an order was placed, the S.O.E. specified that pistols and revolvers would be delivered in H-3 and H-5 containers (for armament and sabotage material), but that "these may be either revolvers or automatics. The latter are in very short supply."

A July 1944 inventory of containers dropped for the FIREMAN network shows that all Colt pistols delivered had been manufactured the previous year in the Remington factories and that most of them came from the same batch. The serial numbers ranged between 1,365,590 and 1,365,988.

The excellent but scarce Canadian Browning pistols, which offered the advantage of using the German Parabellum 9-millimeter cartridge, were exclusively reserved for the special British Commando and Airborne units. They were available to S.O.E. network chiefs who requested them, but it was unusual for a Maquis to have one in his hands. Without a doubt, the representative pistol of the Resistance was the Colt. Fifty cartridges were usually delivered with a weapon.

Pistols and revolvers were primarily meant for self-defense. We now need to look at offensive weapons. In the following pages, we will review submachine guns, the U.S. M-1 paratrooper carbine, and rifles, in order of cartridge power.

# SUBMACHINE GUNS

## Sten Mark II (Tubular Butt)

***Called No. 51 by the S.O.E.***
***Caliber: 9-Millimeter Parabellum***

*Scale: 1/3*

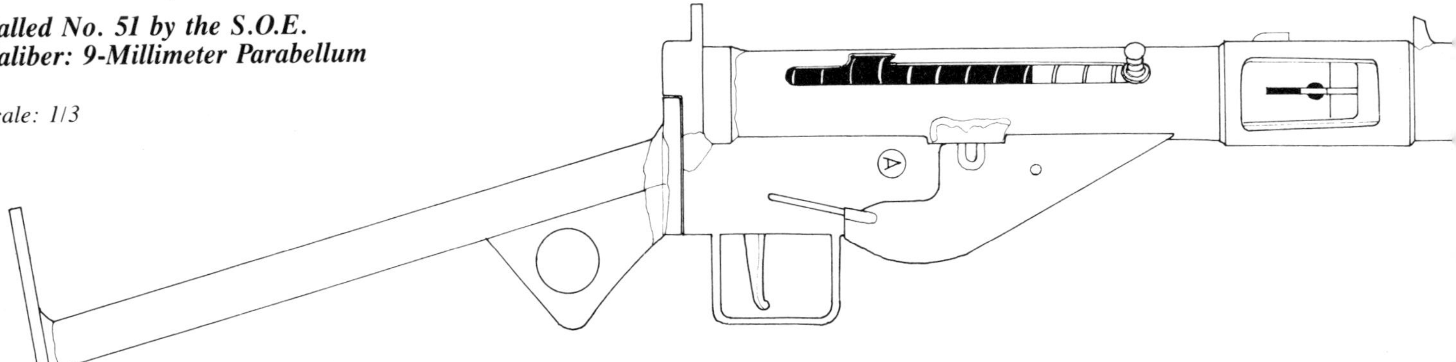


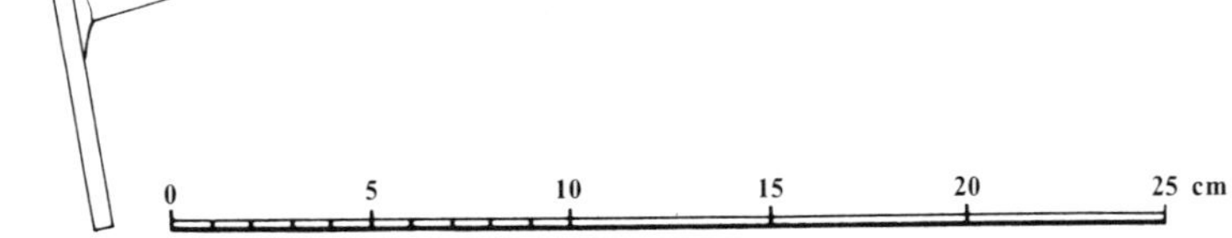




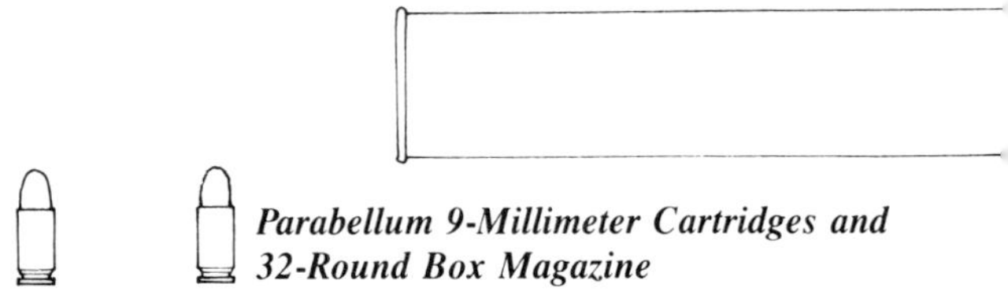

***Parabellum 9-Millimeter Cartridges and 32-Round Box Magazine***

*Scale: 1/3*

## Sten Mark II (Skeleton-Frame Butt)

*Except for the butt, which was made out of a U-shaped and welded piece of iron, this weapon was identical to the one above.*

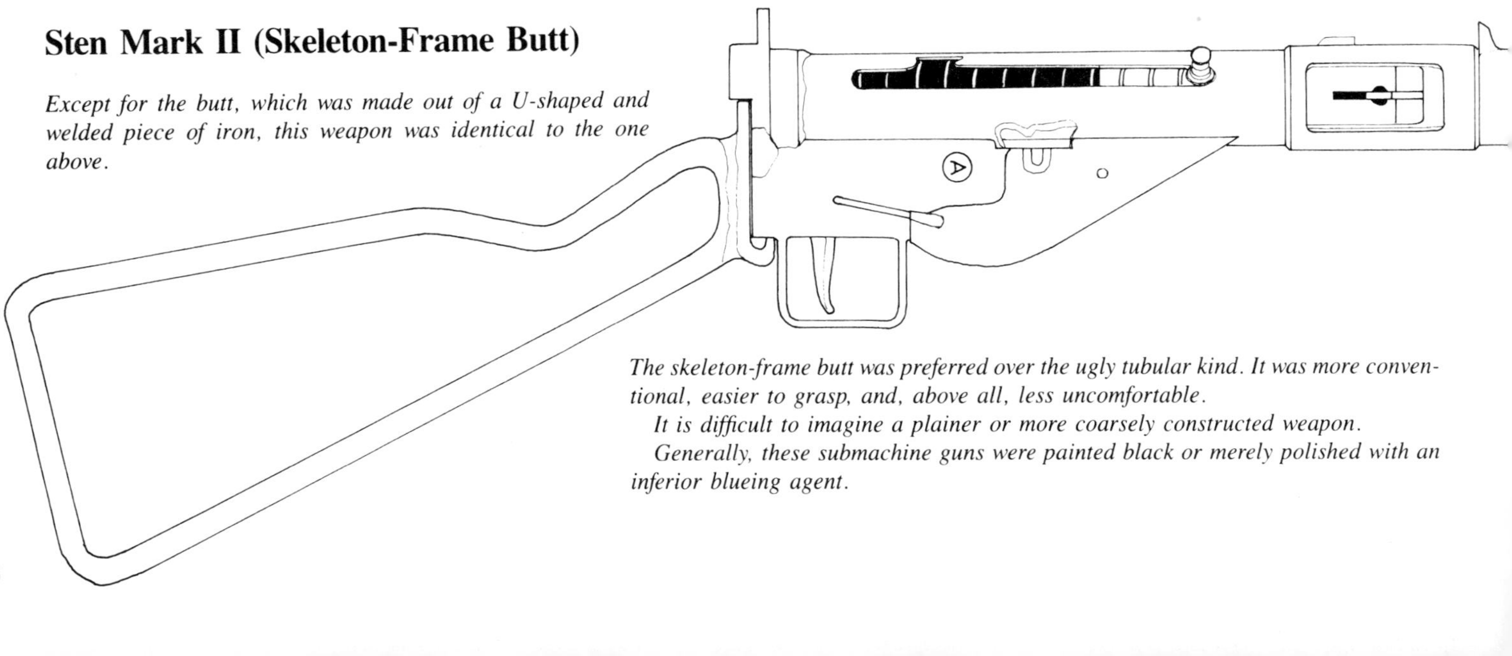


*The skeleton-frame butt was preferred over the ugly tubular kind. It was more conventional, easier to grasp, and, above all, less uncomfortable.*

*It is difficult to imagine a plainer or more coarsely constructed weapon.*

*Generally, these submachine guns were painted black or merely polished with an inferior blueing agent.*

*These desperate-looking firearms, products of a desperate time, were really a brilliant technical improvisation. The Sten could be disassembled into 3 parts and carried in any type of rucksack or concealed in a suitcase. The Sten performed the same functions as the very handsome and very costly German submachine guns. It had a selector button that could be switched over for single-shot firing.*

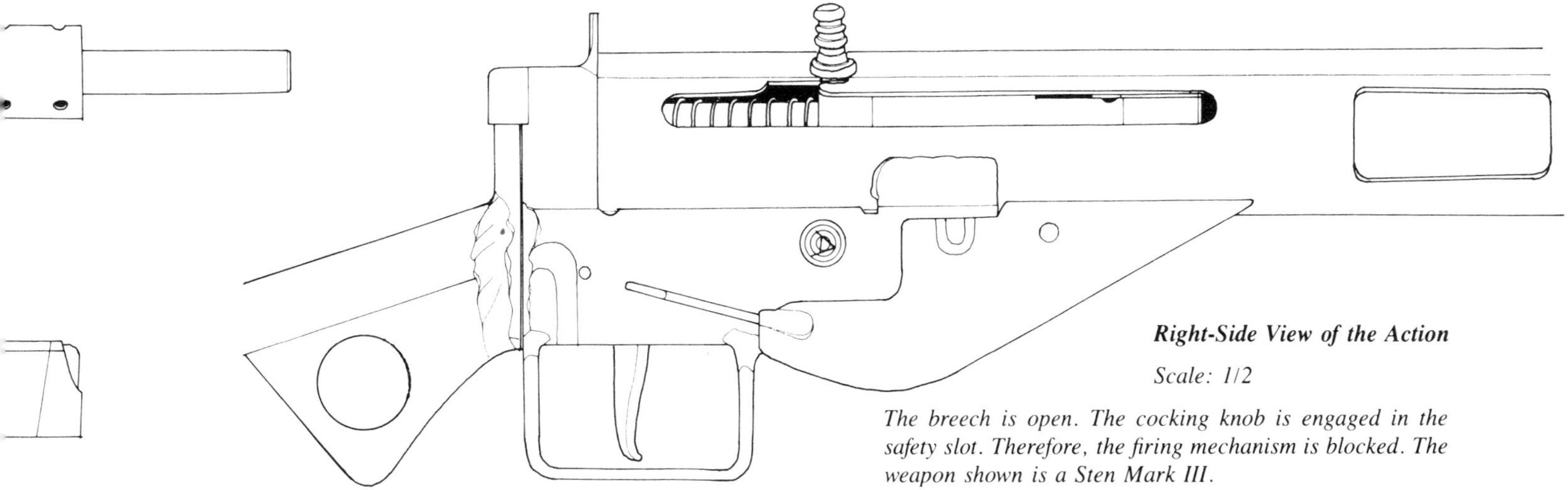

***Right-Side View of the Action***

*Scale: 1/2*

*The breech is open. The cocking knob is engaged in the safety slot. Therefore, the firing mechanism is blocked. The weapon shown is a Sten Mark III.*

***Right-Side View of the Action***

*Scale: 1/2*

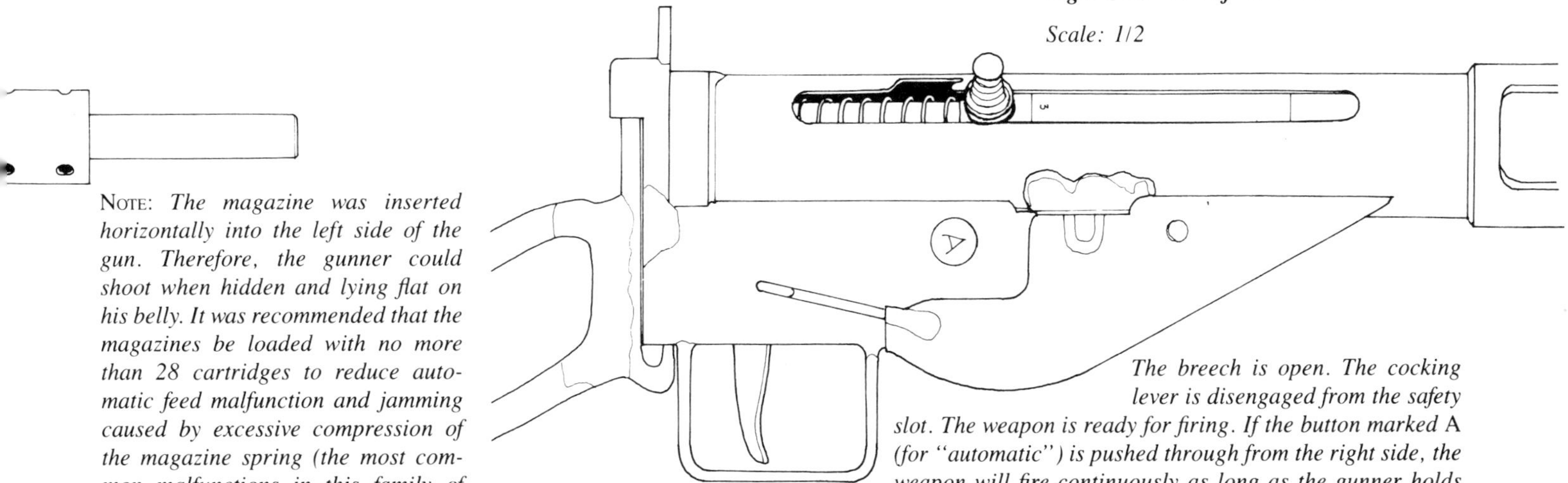


Note: *The magazine was inserted horizontally into the left side of the gun. Therefore, the gunner could shoot when hidden and lying flat on his belly. It was recommended that the magazines be loaded with no more than 28 cartridges to reduce automatic feed malfunction and jamming caused by excessive compression of the magazine spring (the most common malfunctions in this family of weapons).*

*The breech is open. The cocking lever is disengaged from the safety slot. The weapon is ready for firing. If the button marked* A *(for "automatic") is pushed through from the right side, the weapon will fire continuously as long as the gunner holds his finger on the trigger.*

## Sten Mark III

***Caliber: 9-Millimeter Parabellum***

*Scale: 1/3*

*The Mark III had a tubular butt identical to that of the Mark II. Guns of this model were usually painted black.*

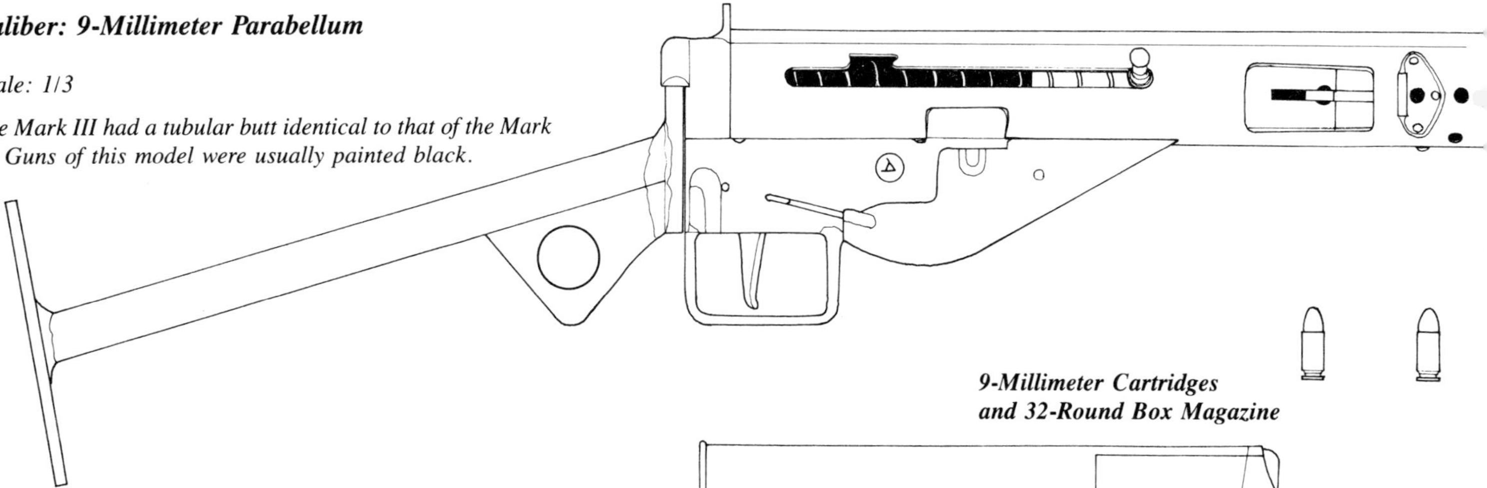

***9-Millimeter Cartridges and 32-Round Box Magazine***

## Sten Mark V

*Equipped with a No. 4 Mark II rifle "spike" bayonet, butt stock, and pistol grip of wood.*

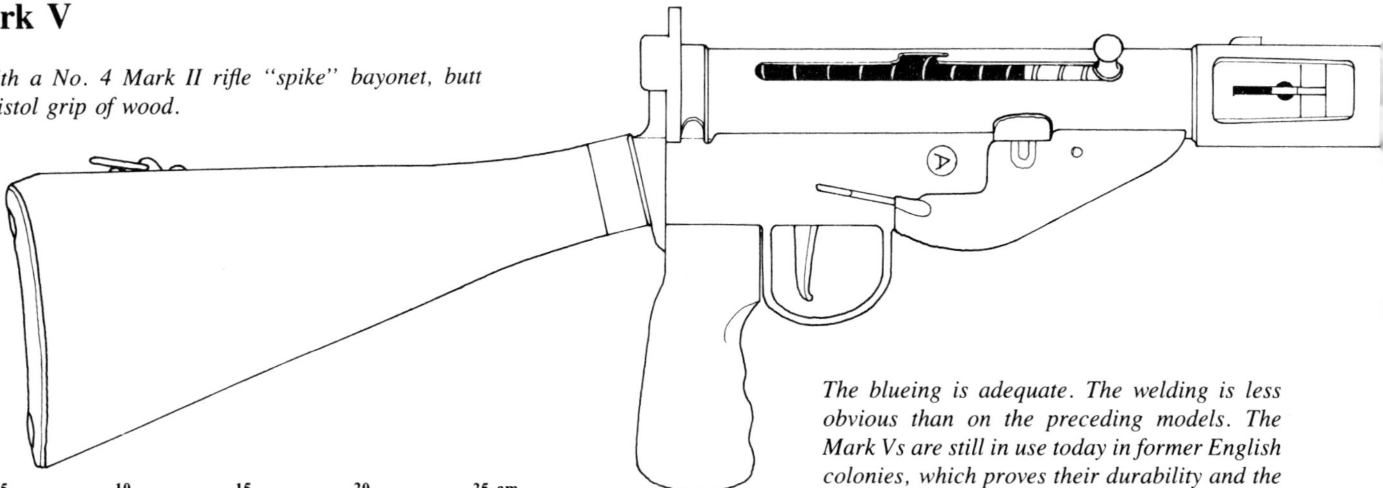

0 5 10 15 20 25 cm

*The blueing is adequate. The welding is less obvious than on the preceding models. The Mark Vs are still in use today in former English colonies, which proves their durability and the worth of their design.*

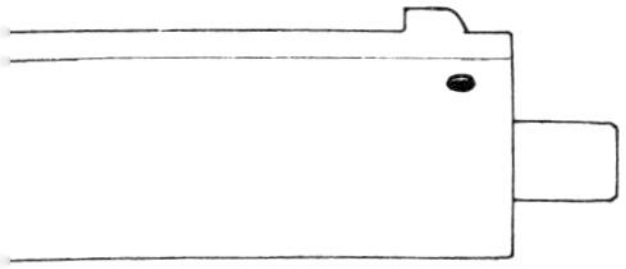

Note *on the Sten Mark III: This model can only be disassembled into 2 parts (the barrel is not detachable as in other models). However, it is easy to grasp. One can hold it tightly while firing, without risking burns or other injuries. Except for the barrel jacket of protective steel sheet tubing surrounding the barrel to the muzzle, this gun is identical to the models preceding it. It was usually equipped with a webbed sling and delivered with 8 loaded magazines.*

*The Mark III was not as widely employed as the Mark II, more of which were dropped by parachute than any other submachine gun.*

***Left-Side View, Butt Removed***

*Scale: 1/3*

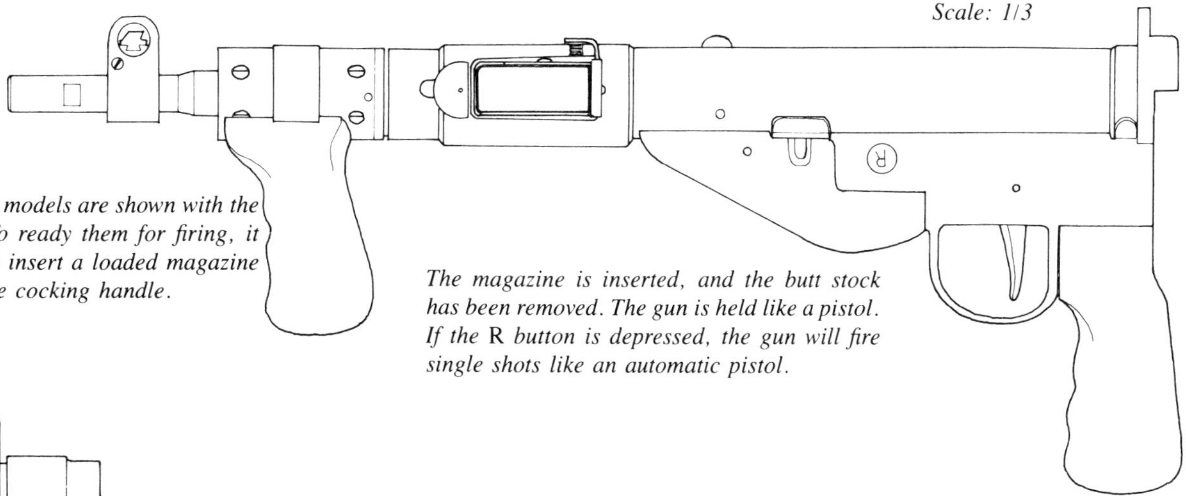



Note: *These Sten models are shown with the breech closed. To ready them for firing, it was necessary to insert a loaded magazine and pull back the cocking handle.*

*The magazine is inserted, and the butt stock has been removed. The gun is held like a pistol. If the* R *button is depressed, the gun will fire single shots like an automatic pistol.*



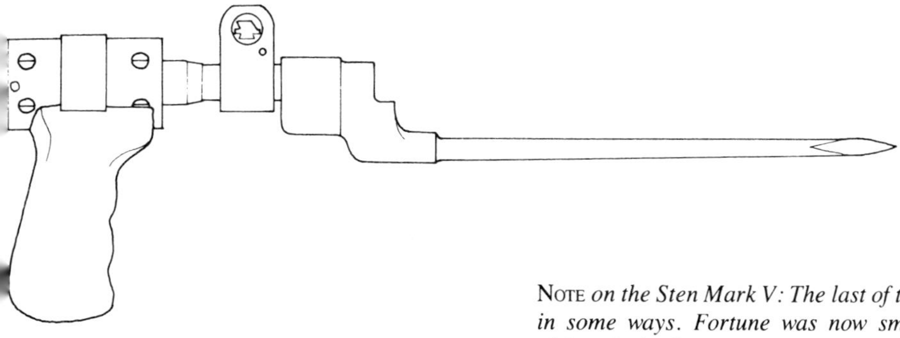

Note *on the Sten Mark V: The last of the Sten models, the Mark V can be considered the deluxe model in some ways. Fortune was now smiling on the Allied forces, and the weapon became refined, although the firing mechanism was identical to that of the earliest Mark IIs. The 2 pistol grips were remarkably easy to hold. The weapon could be easily used without a butt and, in this way, became a veritable "machine-pistol." The barrel was carefully machine-finished, carried an adjustable front sight and accepted the "spike" bayonet. These weapons were especially used by British paratroopers at Arnhem. They appeared rather late in the war, and only a few were dropped by parachute for the Resistance fighters.*

# Thompson "Tommy Gun" 1921 Model

***Caliber: .45***

*Scale: 1/3*

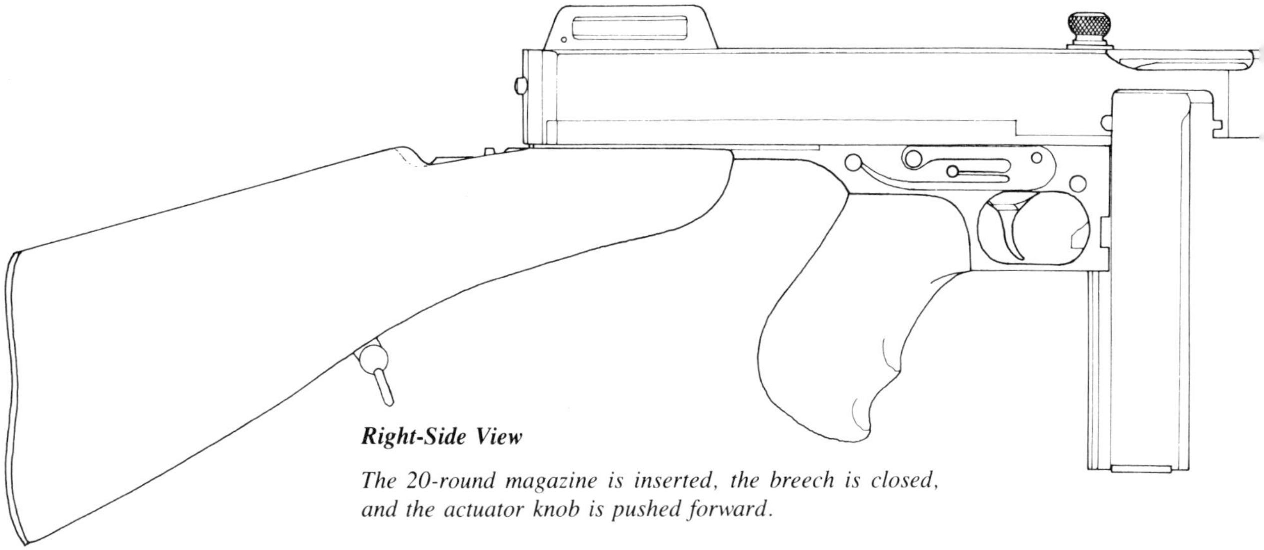

***Right-Side View***

*The 20-round magazine is inserted, the breech is closed, and the actuator knob is pushed forward.*

# Thompson "Tommy Gun" 1928 Model

*The barrel is equipped with a Cutts compensator or muzzle brake.*

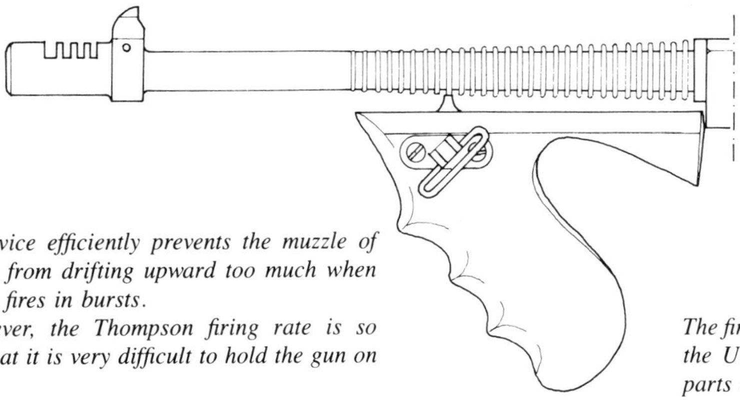

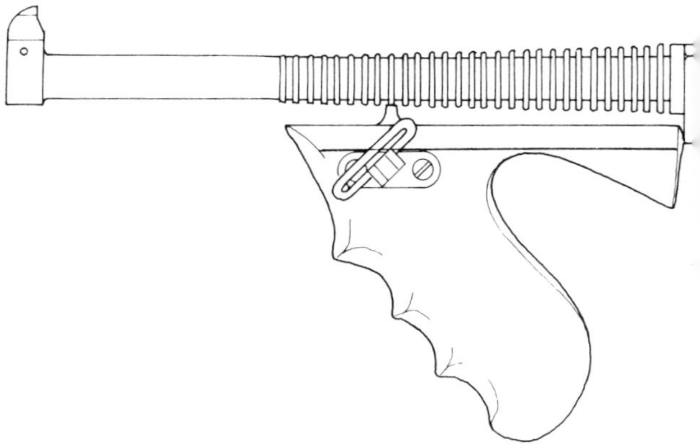

*This device efficiently prevents the muzzle of the gun from drifting upward too much when the gun fires in bursts.*

*However, the Thompson firing rate is so great that it is very difficult to hold the gun on target.*

*The finish of these guns, which were designed and manufactured in the U.S. during peacetime, is of exceptional quality. All metal parts are carefully blued.*

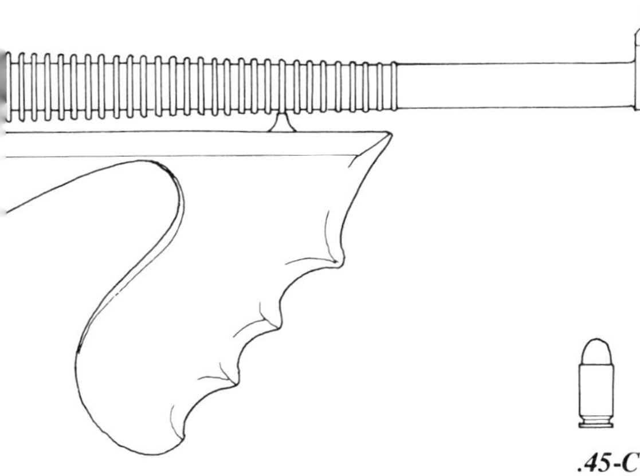

NOTE: *The submachine gun shown here is equipped with a box magazine holding twenty .45 cartridges. The cartridges are the same ones as those used in the regulation Colt automatic.*

*There is also a round, 50-cartridge drum magazine. These guns, with drum-shaped magazines, were delivered to the French and British forces before the war. A great many were used during the first two years of the war, most notably as regulation armament for the famous Commandos.*

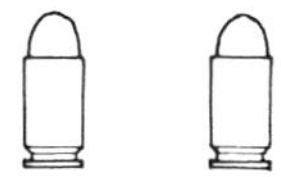

***.45-Caliber Cartridges***

*Scale: 1/3*

***Left-Side View***

*The magazine is inserted. The 4 small holes allow the user to determine the number of cartridges inserted. There is a fire selector switch, set here for full automatic fire, a safety catch, and a magazine catch. The gun shown here will be ready for firing after the actuator knob is pulled back.*

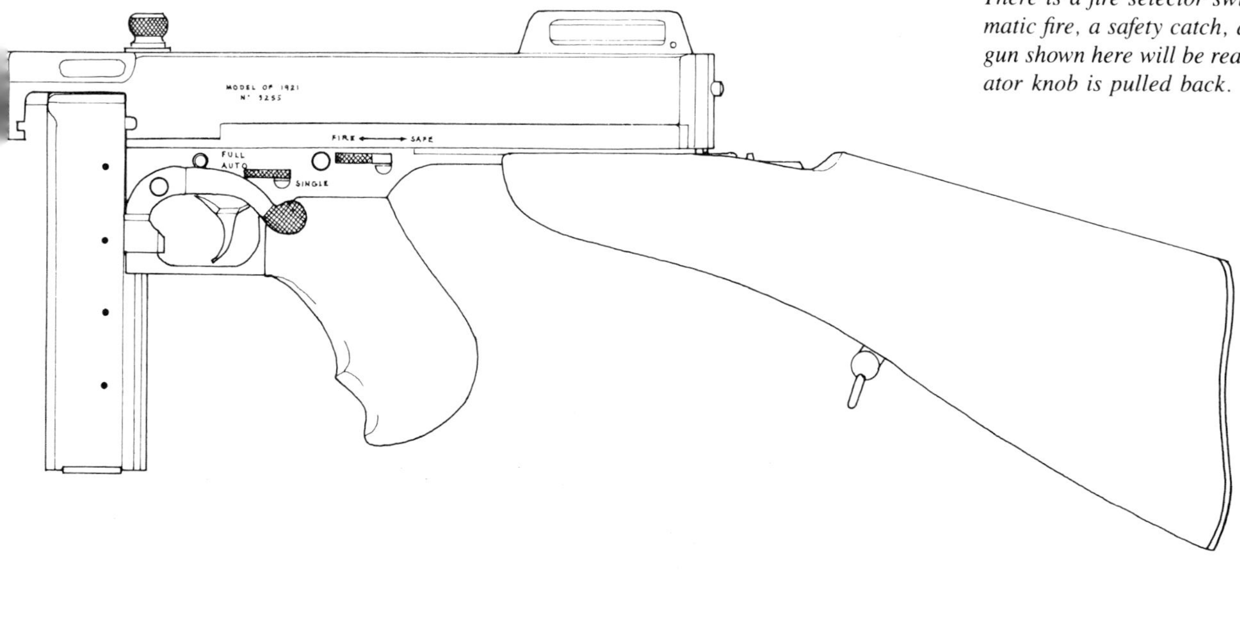

## Thompson "Tommy Gun" Model M-1 (1942)

***Caliber: .45***

*Scale: 1/3*

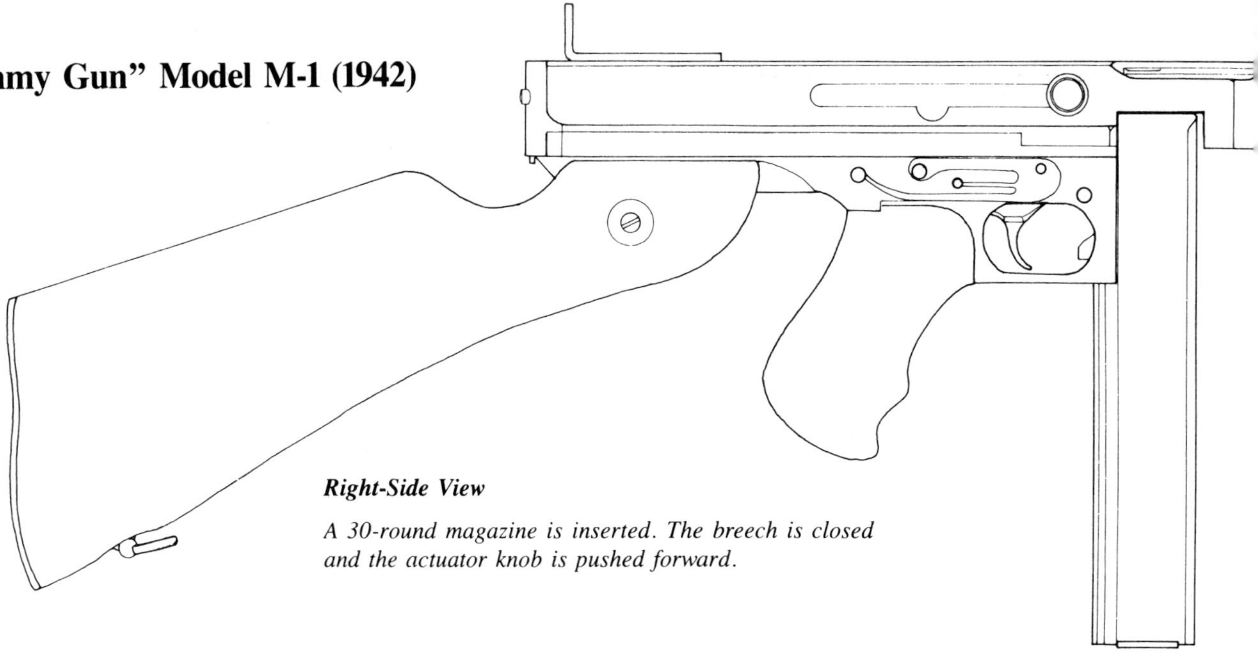

***Right-Side View***

*A 30-round magazine is inserted. The breech is closed and the actuator knob is pushed forward.*

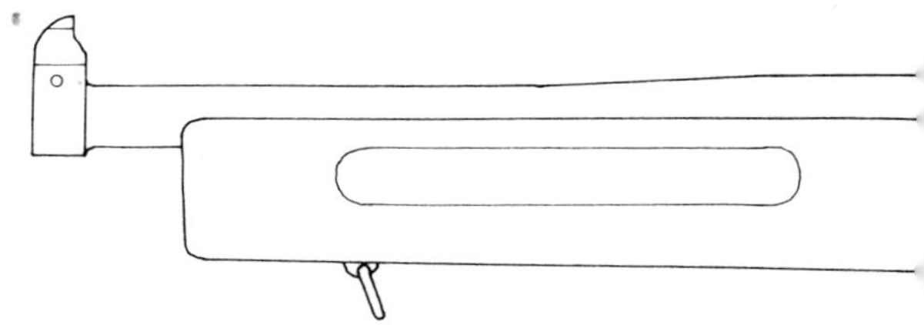

## Thompson "Tommy Gun" 1928 Model A-1

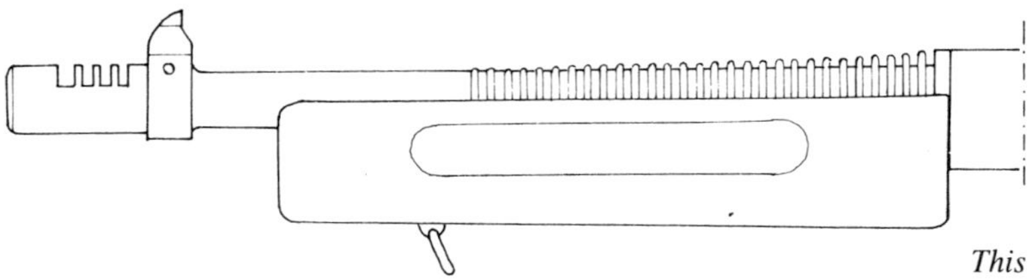

*This model, adopted by the U.S. army, had a Cutts compensator on the muzzle. The initial models kept the radial cooling fins, but these were eventually omitted. The front pistol grip has been replaced by a horizontal foregrip. The actuator knob is still situated on top of the receiver.*

*The French and British governments bought the 1921 and 1928 models before hostilities broke out.*

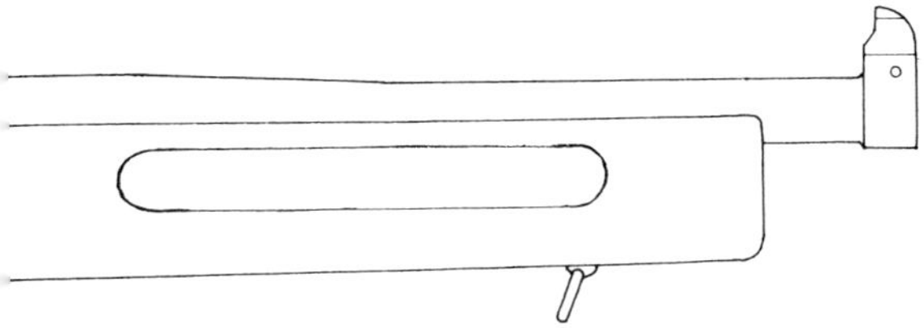

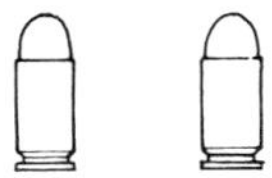

**.45-Caliber Cartridges**

*Scale: 1/3*

NOTE: *The M-1 model, manufactured when the United States entered the war, was a simplified version of the earlier models.*

*The actuator knob found its logical place on the side of the weapon. Box magazines were the only kind employed, containing 20 or 30 cartridges. There was an* L *sight for firing at 100 to 200 yards, or a single aperture on some models, such as the one pictured. The action was simplified from delayed blowback to simple blowback. The butt stock could no longer be removed, the appearance was simpler, and the finish was that of a wartime product. The production costs had dropped considerably, but the weapon was still too heavy and cumbersome for clandestine operations.*

*Nevertheless, compared to the Sten, the Thompsons were highly sought after because of the high quality of their manufacture and their dependability.*

*Only a very small number of these guns were dropped by parachute.*

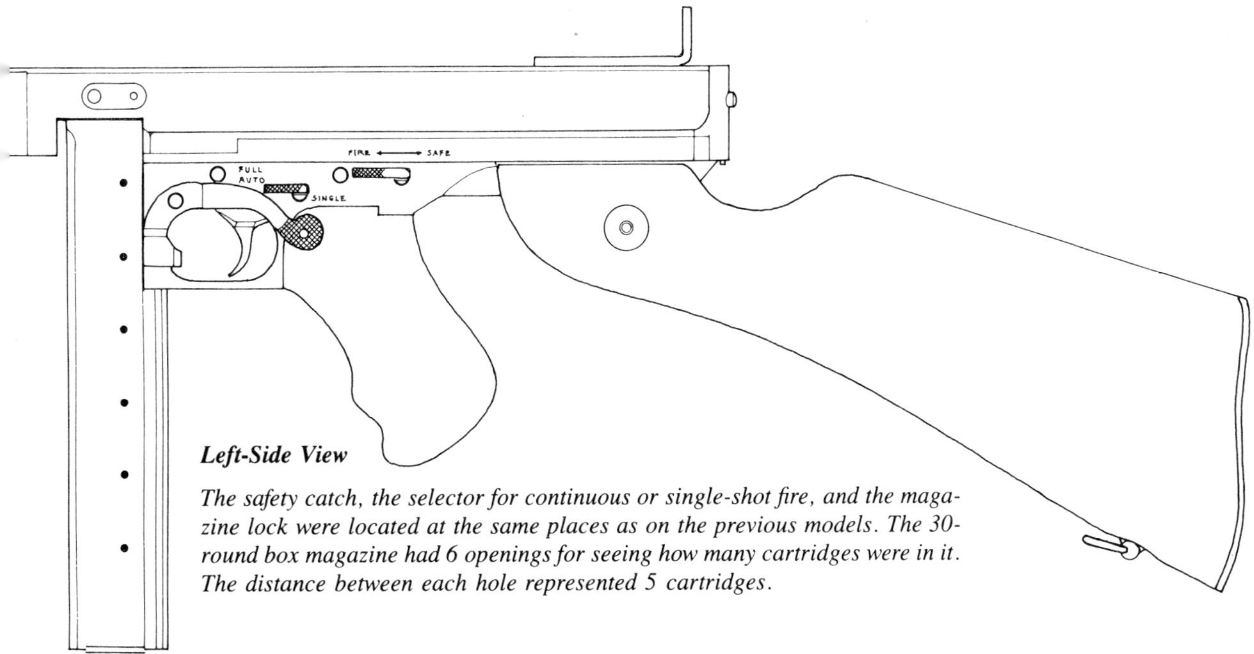


***Left-Side View***

*The safety catch, the selector for continuous or single-shot fire, and the magazine lock were located at the same places as on the previous models. The 30-round box magazine had 6 openings for seeing how many cartridges were in it. The distance between each hole represented 5 cartridges.*

# U.S. M-3 (1943)

***Calibers: .45 or 9-Millimeter Parabellum (By Changing the Barrel)***

*Scale: 1/3*

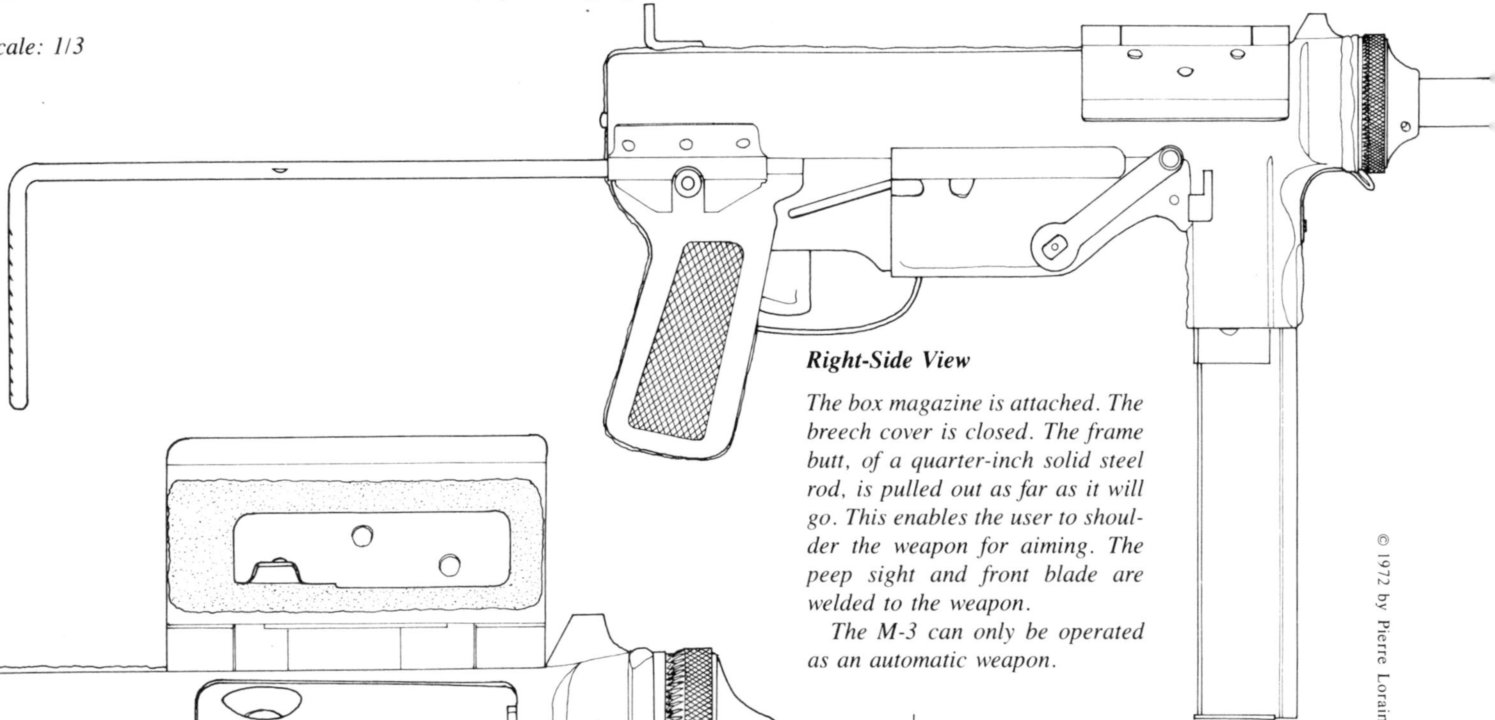

***Right-Side View***

*The box magazine is attached. The breech cover is closed. The frame butt, of a quarter-inch solid steel rod, is pulled out as far as it will go. This enables the user to shoulder the weapon for aiming. The peep sight and front blade are welded to the weapon.*

*The M-3 can only be operated as an automatic weapon.*



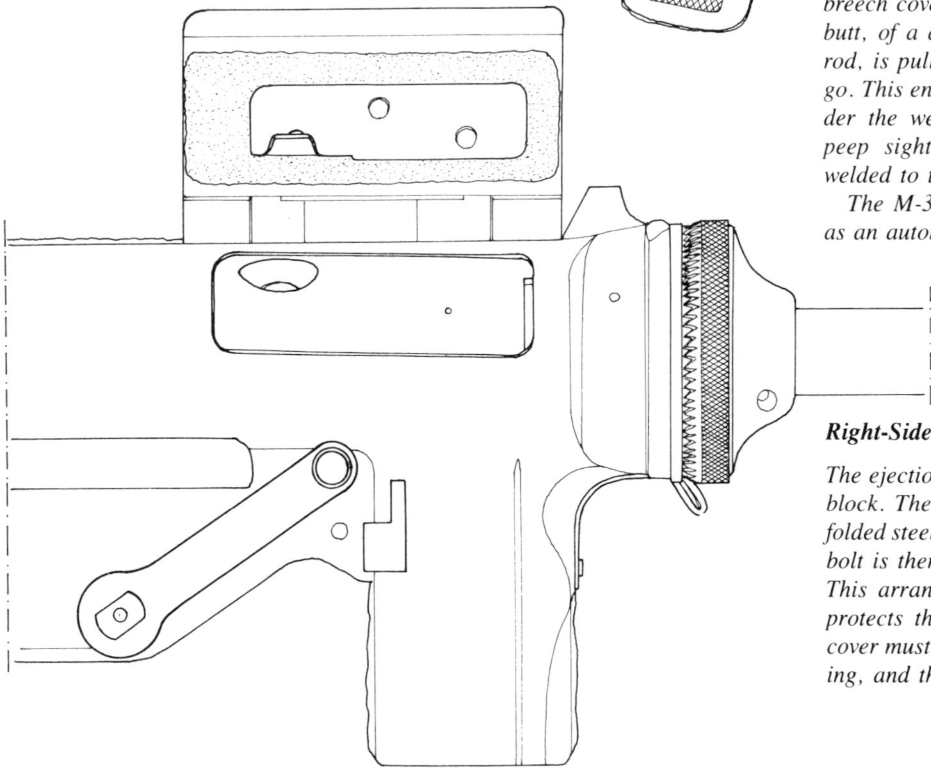

***Right-Side View of the Breech***

*The ejection port cover is partially open, revealing the bolt block. The latter has a hole drilled in it, and a tongue of folded steel lodges in it when the breech cover is closed. The bolt is then immobilized, and the weapon cannot be fired. This arrangement fills the role of a safety lock and also protects the mechanism from dirt. For firing, the breech cover must be fully opened, a magazine inserted in its housing, and the cocking lever pulled back.*

NOTE: *Except for the bolt and the barrel, the M-3 was completely manufactured out of sheet steel stamped and welded. The body is made of 2 stamped sheet-steel symmetrical half shells soldered together at the edges. The result was a rough and ugly but safe and simple machine that looks more like a tool (an electric drill) than a weapon.*

*The M-3 was completely painted olive green.*

*The weapon represented the U.S. solution to the demand for rapid and intensive weapon production during times of crisis. It was the homologue of the British Sten.*

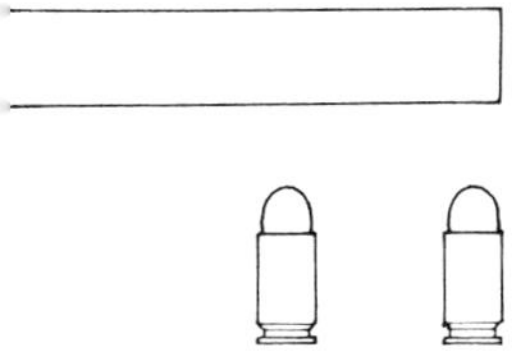

***.45-Caliber Cartridges***

*Scale: 1/3*

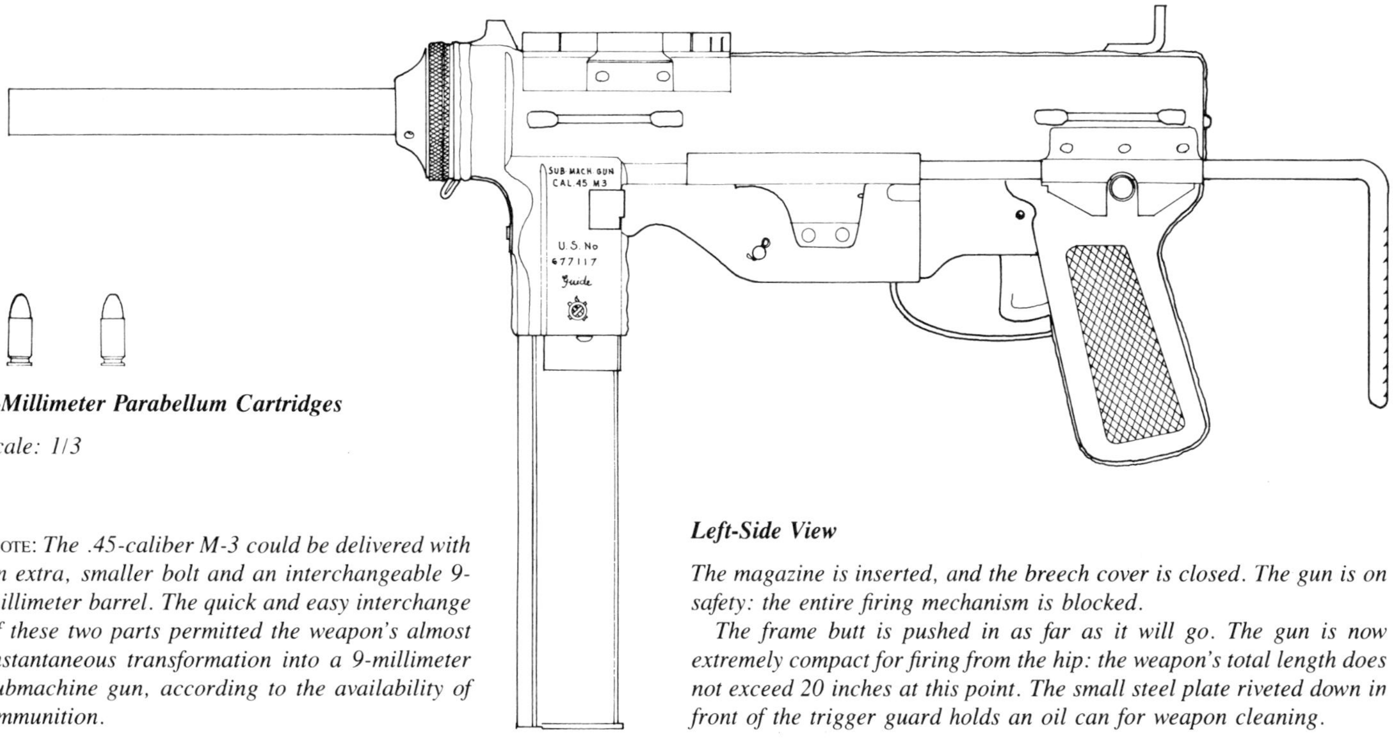


***9-Millimeter Parabellum Cartridges***

*Scale: 1/3*

NOTE: *The .45-caliber M-3 could be delivered with an extra, smaller bolt and an interchangeable 9-millimeter barrel. The quick and easy interchange of these two parts permitted the weapon's almost instantaneous transformation into a 9-millimeter submachine gun, according to the availability of ammunition.*

***Left-Side View***

*The magazine is inserted, and the breech cover is closed. The gun is on safety: the entire firing mechanism is blocked.*

*The frame butt is pushed in as far as it will go. The gun is now extremely compact for firing from the hip: the weapon's total length does not exceed 20 inches at this point. The small steel plate riveted down in front of the trigger guard holds an oil can for weapon cleaning.*

# CARBINE

## Winchester Semiautomatic Carbine U.S. M-1 1941 Model

***Caliber: .30 M-1***

*Scale: 1/3*

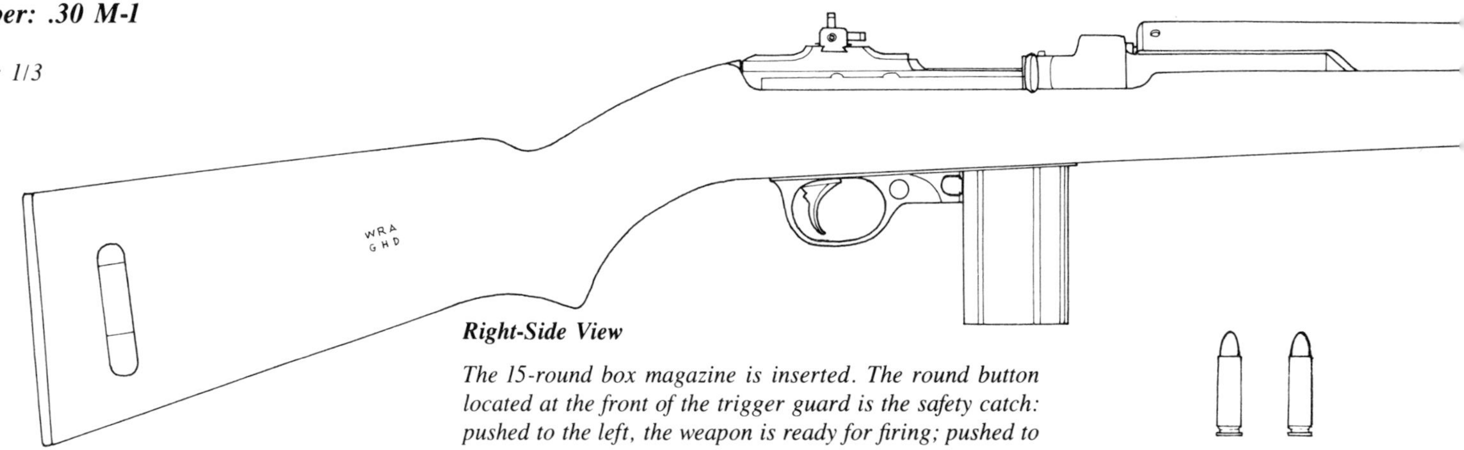


***Right-Side View***

*The 15-round box magazine is inserted. The round button located at the front of the trigger guard is the safety catch: pushed to the left, the weapon is ready for firing; pushed to the right, the trigger is blocked and the gun cannot fire. The other button, in front of the safety lock, is the magazine lock, which locks the magazine securely in the trigger housing. The initials* W.R.A. *on the stock of the carbine indicate that it is manufactured by Winchester Repeating Arms.*

***.30 M-1 Cartridges***

*Scale: 1/2*

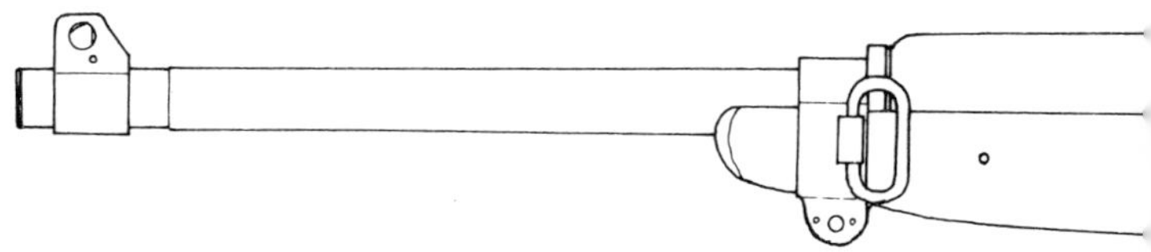

NOTE: *The M-1 carbine cannot maintain shoot-by-burst fire, but fires only one shot at a time. The action of the finger on the trigger causes, in succession, the ignition of the cartridge, the expulsion of the bullet, the extraction and the ejection of the spent cartridge case, and the introduction of a new cartridge into the chamber.*

*Firing is rapid and reliable: jamming does not often occur.*

STANDARD FEATURES

*Overall length: 35.65 inches*
*Barrel length: 18 inches*
*Weight unloaded: 5 pounds, 7 ounces*
*Action: gas, semiautomatic*
*Magazine capacity: 15 or 30 rounds*
*Muzzle velocity: 1,970 feet per second*
*Muzzle energy: 950 foot-pounds*
*Maximum useful range: 300 yards*

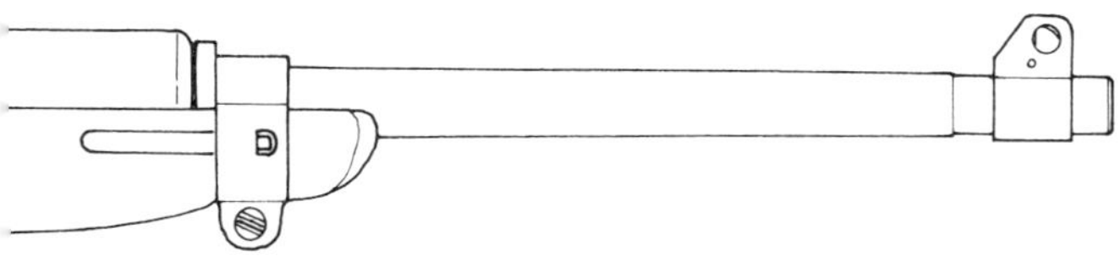

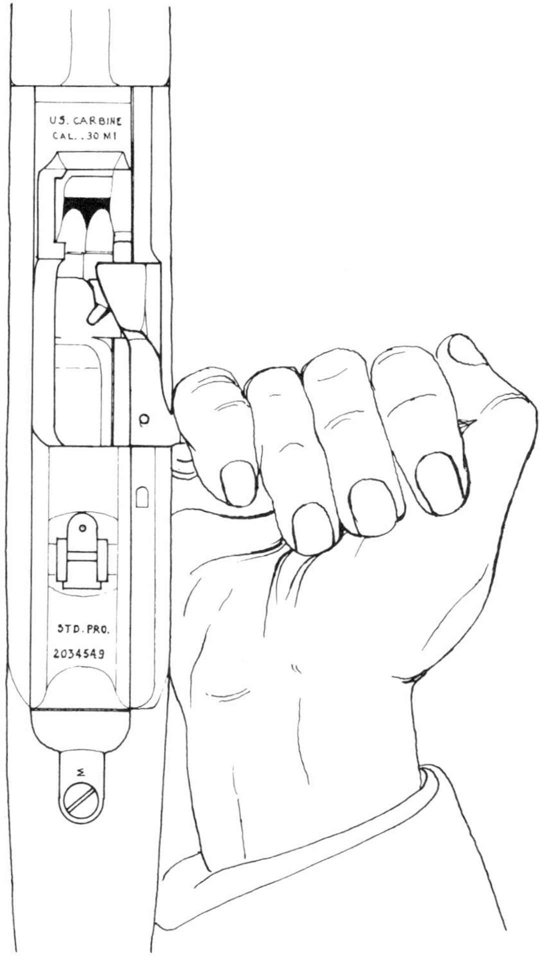


***Detail, Viewed from Above***

*Scale: 1/2*

*A magazine has just been inserted from below. To load the weapon, all that is needed is to draw back the operating slide handle. This action unlocks the bolt and draws it backward. If the operating slide handle is released at the end of its course, the bolt returns forward, pushed by the return spring, drawing the operating slide mechanism which pushes the first cartridge into the chamber. At the end of its course, the bolt pivots to the right and locks securely. All that is needed now is to squeeze on the trigger for each shot.*

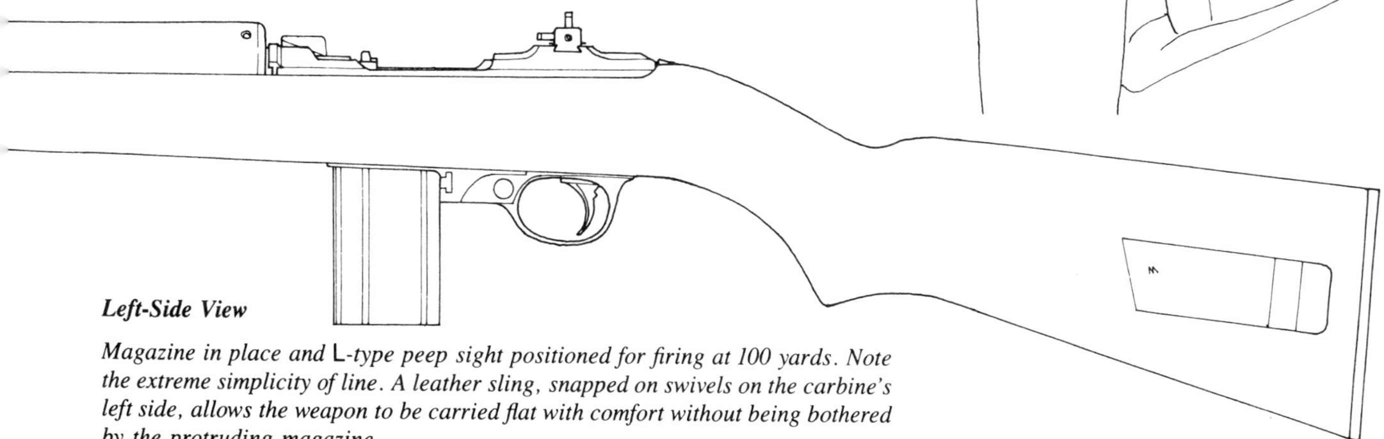

***Left-Side View***

*Magazine in place and L-type peep sight positioned for firing at 100 yards. Note the extreme simplicity of line. A leather sling, snapped on swivels on the carbine's left side, allows the weapon to be carried flat with comfort without being bothered by the protruding magazine.*

# RIFLES

## Lee-Enfield No. 1 S.M.L.E. Mark III* 1916 Model

*(S.M.L.E. = Short Magazine Lee-Enfield)*

***Caliber: .303***

*Scale: 1/3*

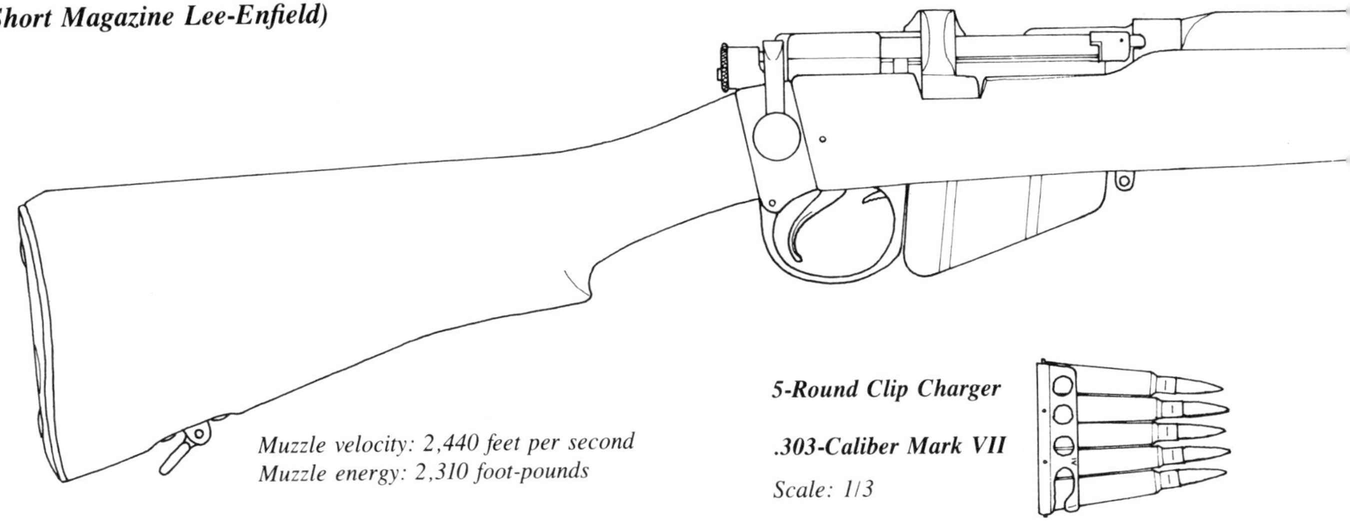

*Muzzle velocity: 2,440 feet per second*
*Muzzle energy: 2,310 foot-pounds*

***5-Round Clip Charger***

***.303-Caliber Mark VII***

*Scale: 1/3*

## Lee-Enfield No. 4 Mark I 1939 Model

***Caliber: .303 British***

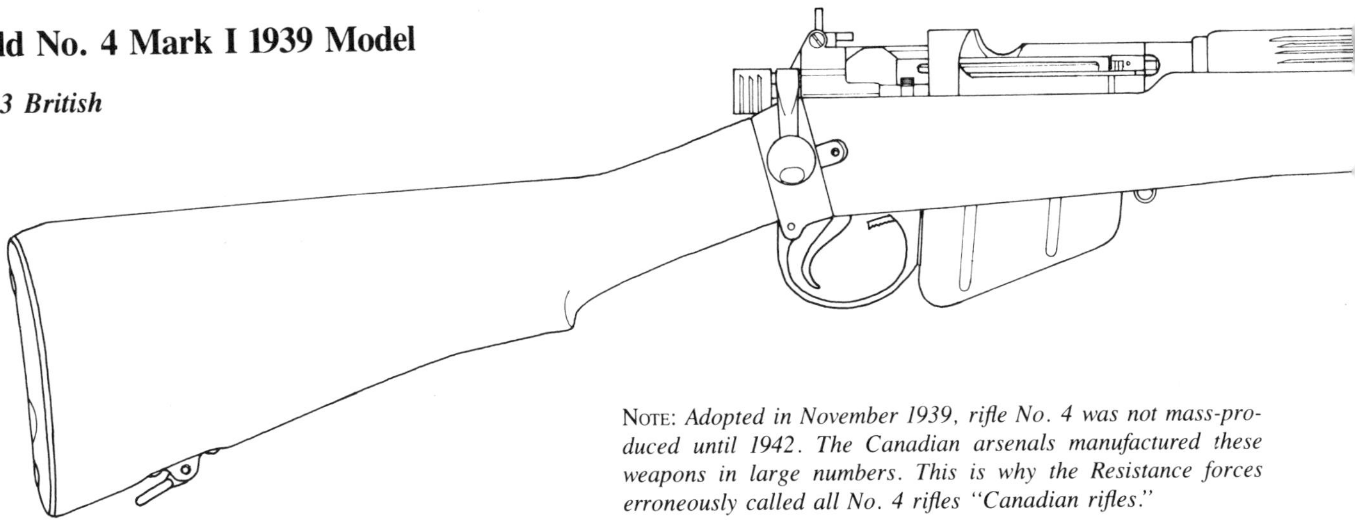



NOTE: *Adopted in November 1939, rifle No. 4 was not mass-produced until 1942. The Canadian arsenals manufactured these weapons in large numbers. This is why the Resistance forces erroneously called all No. 4 rifles "Canadian rifles."*

*Overall length of each rifle: 44.5 inches*
*Length of each barrel: 25.19 inches*
*Weight of the Mark III*: 8 pounds, 2 ounces*
*Weight of No. 4: 8 pounds, 13 ounces*

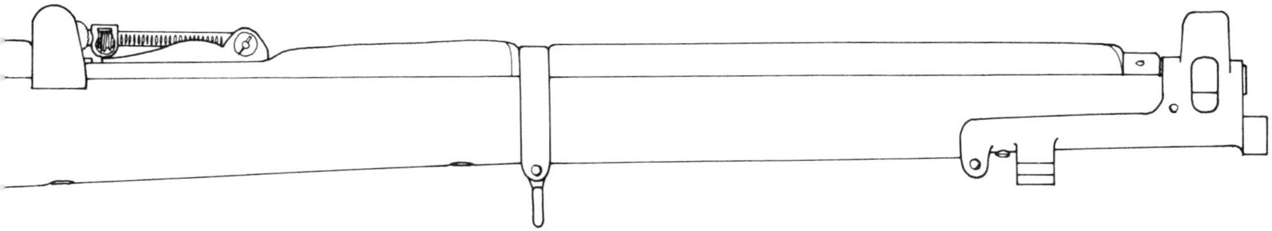

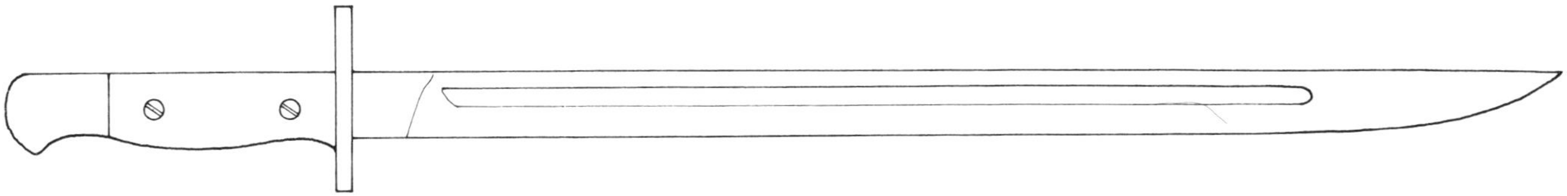

***No. 1 Mark I Bayonet***

*Overall length: 22 inches*

NOTE: *The magazine capacity for both weapons is 2 clips or 10 rounds.*

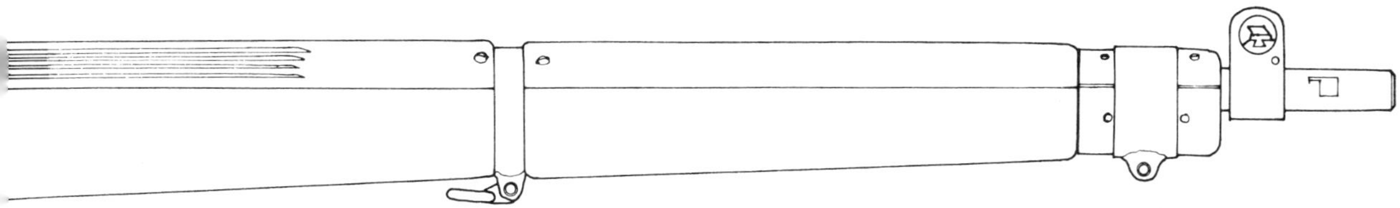

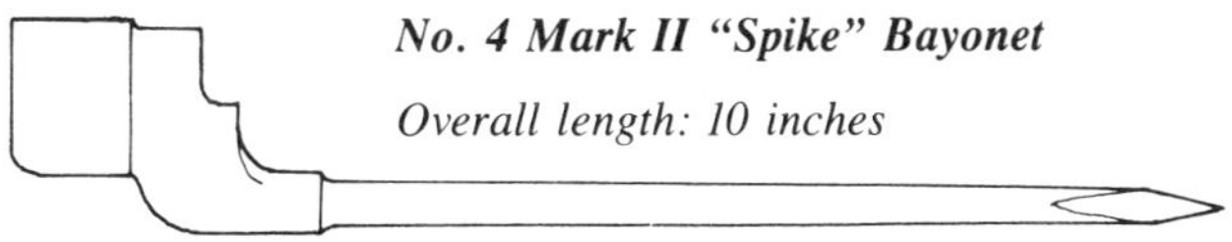

***No. 4 Mark II "Spike" Bayonet***

*Overall length: 10 inches*

*This rifle was really the regulation arm for Britain and her Commonwealth from 1939 to 1957. Australia, however, never adopted the No. 4 rifle and continued to manufacture the Mark III*.*

## *Details of Lee-Enfield No. 1 S.M.L.E. Mark III* 1916 Model*

*Scale: 1/2*



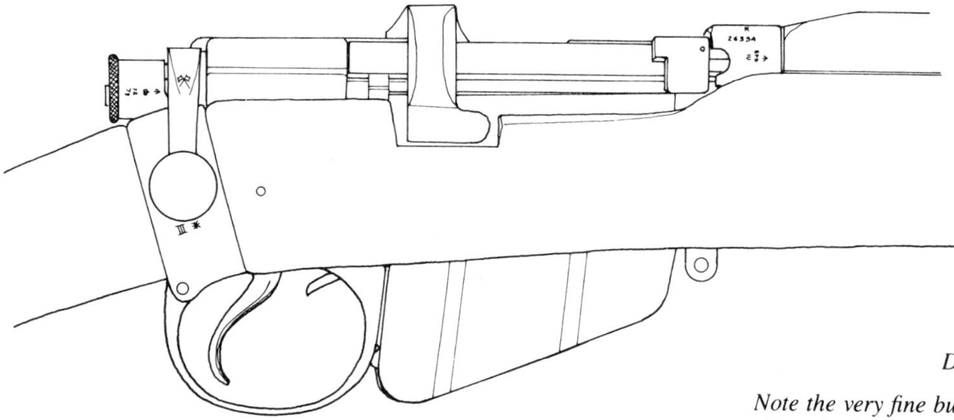

### *Right-Side View of the Lock*

*The bolt is closed and locked. The hammer is uncocked.*

*The magazine in front of the trigger guard is removable; its catch is located inside the trigger guard, in front of the trigger.*

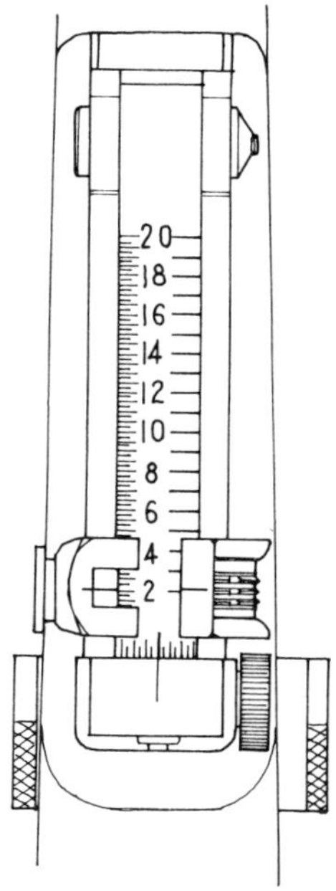


### *Rear sight*

*Detail: Scale: 2/3*

*Note the very fine but very costly precision leaf sight with drift adjustment by a micrometer screw controlled by a serrated click knob.*

### *Left-Side View of the Lock*

*The bolt is closed and locked. The hammer is uncocked. The safety is not engaged.*

*To put the weapon on the safety, the rifleman must pull the thumb lever marked* EFD *(Enfield) straight back. This lever is next to the hammer.*

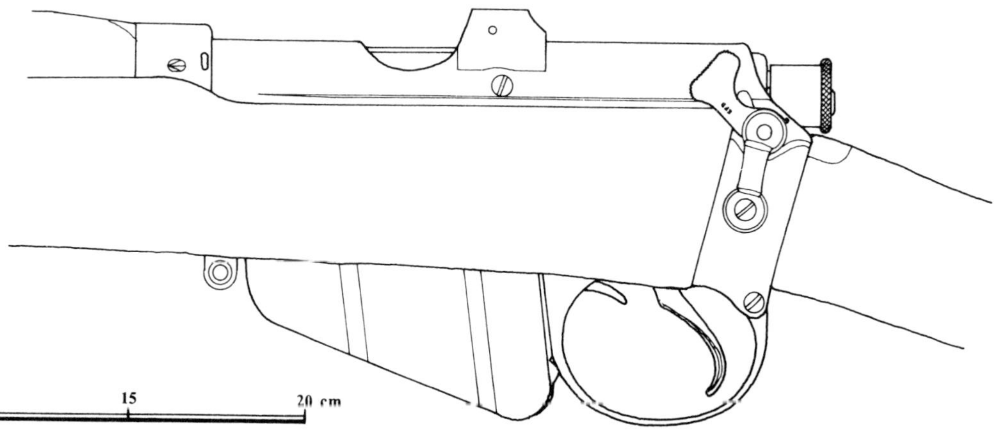

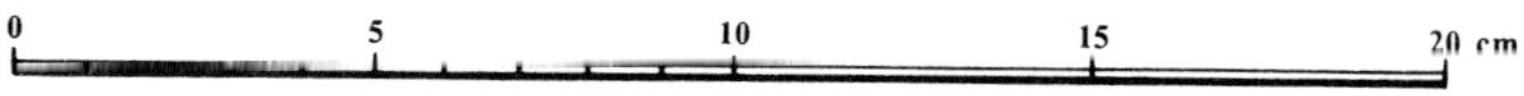

## *Details of Lee-Enfield No. 4 Mark I 1939 Model*

*Scale: 1/2*

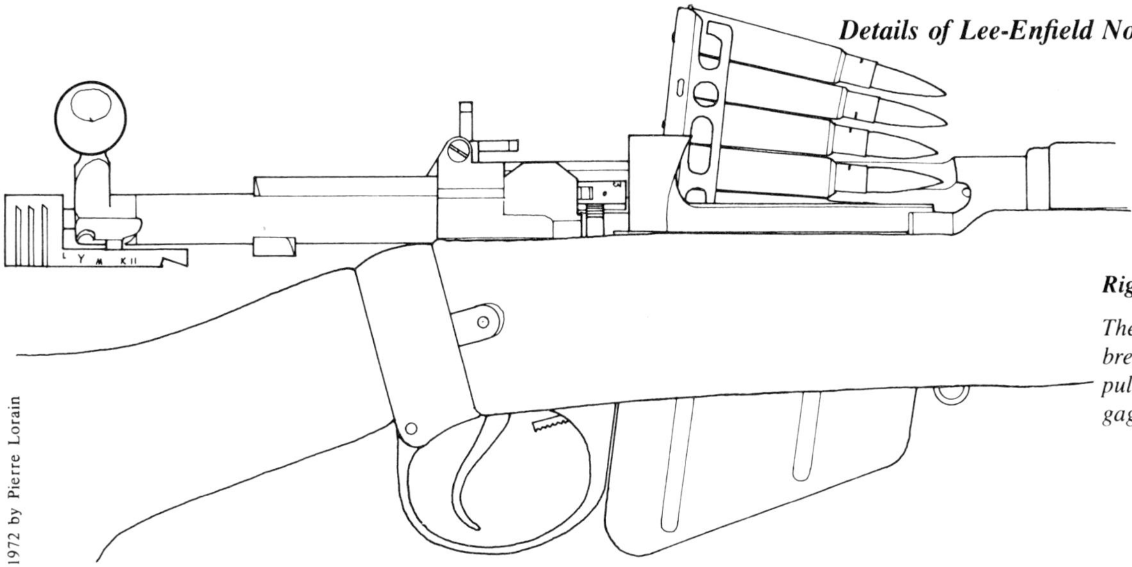

**Right-Side View**

*The rifle is being loaded. The breech is open and the bolt pulled back. The clip is engaged in the clip guide.*



NOTE: *The rifle pictured is a simplified breakdown* L*-type rear sight with 2 battle peep sights set for 300 and 600 yards.*

**Another Type of Rear Sight**

*Detail from above; scale: 2/3*

*This sight was adopted in 1943. It could be adjusted to any range between 200 and 1,300 yards.*

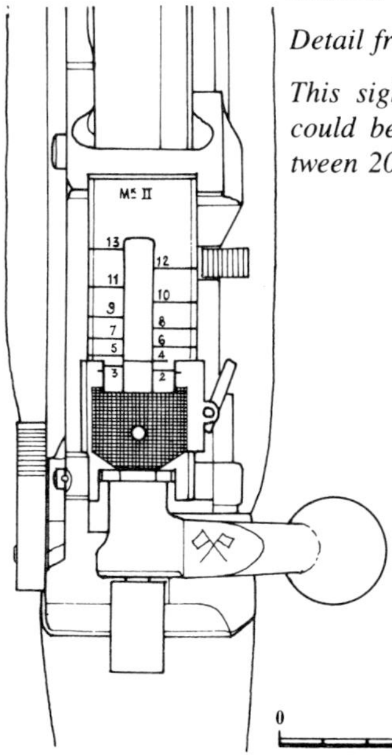


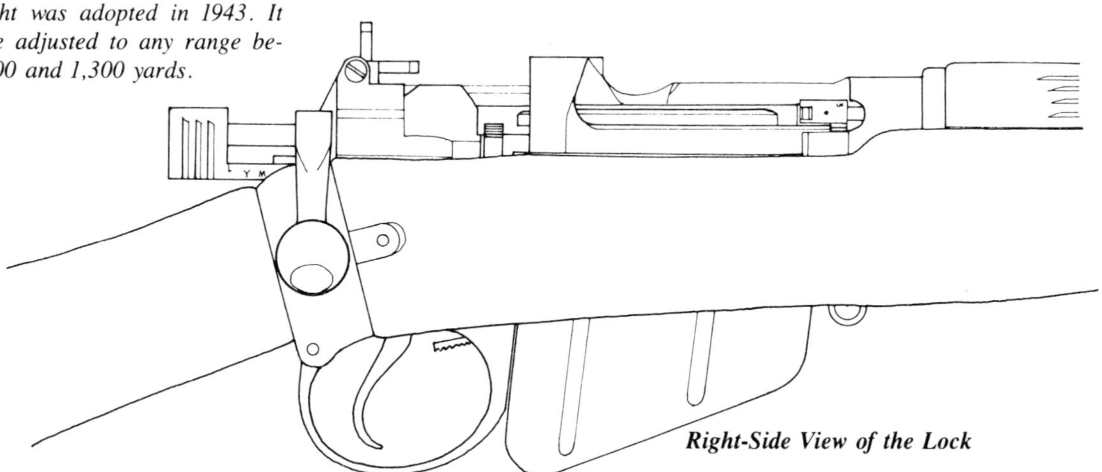

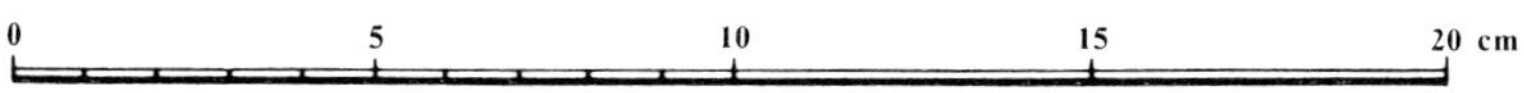


**Right-Side View of the Lock**

*The bolt is closed and locked. The striker is cocked. The weapon is ready to be fired.*

### *Details of Lee-Enfield No. 4 Mark I 1939 Model*

*Weapon being loaded. The clip is engaged in its guides. Thumb pressure is going to slide the 5 cartridges into the magazine. The contents of 2 clips could thus be inserted, which gave the English weapon double the capacity of most weapons then in use, including the German rifle.*

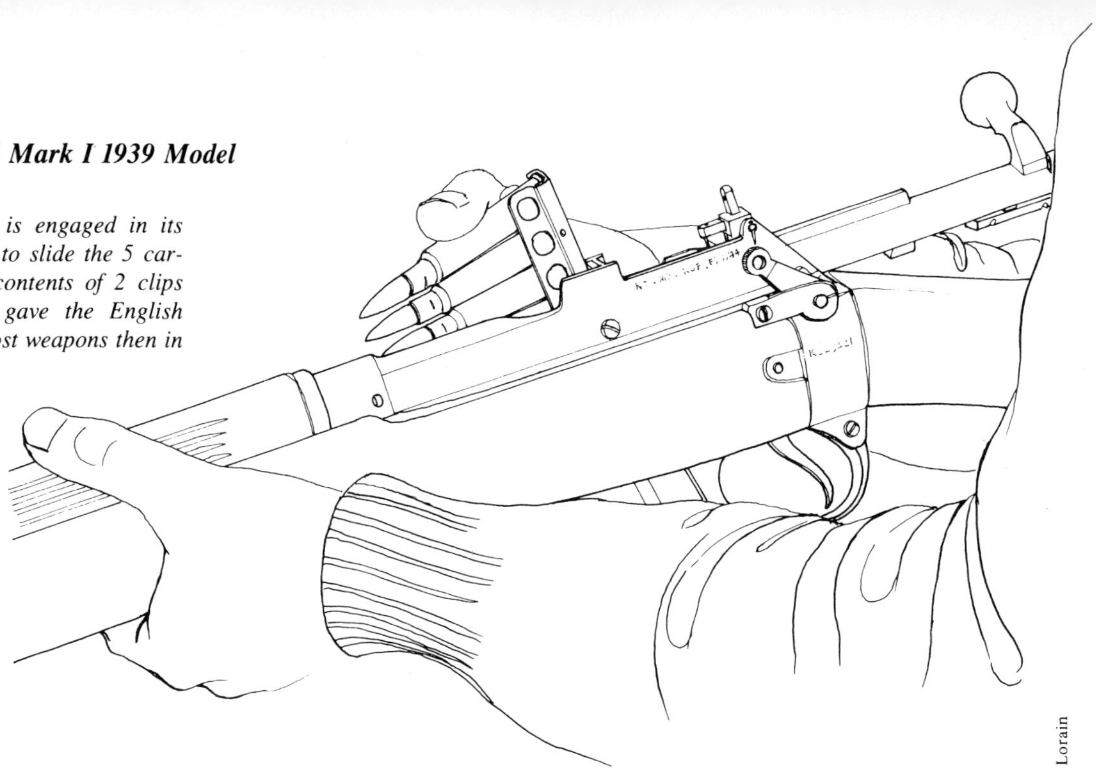

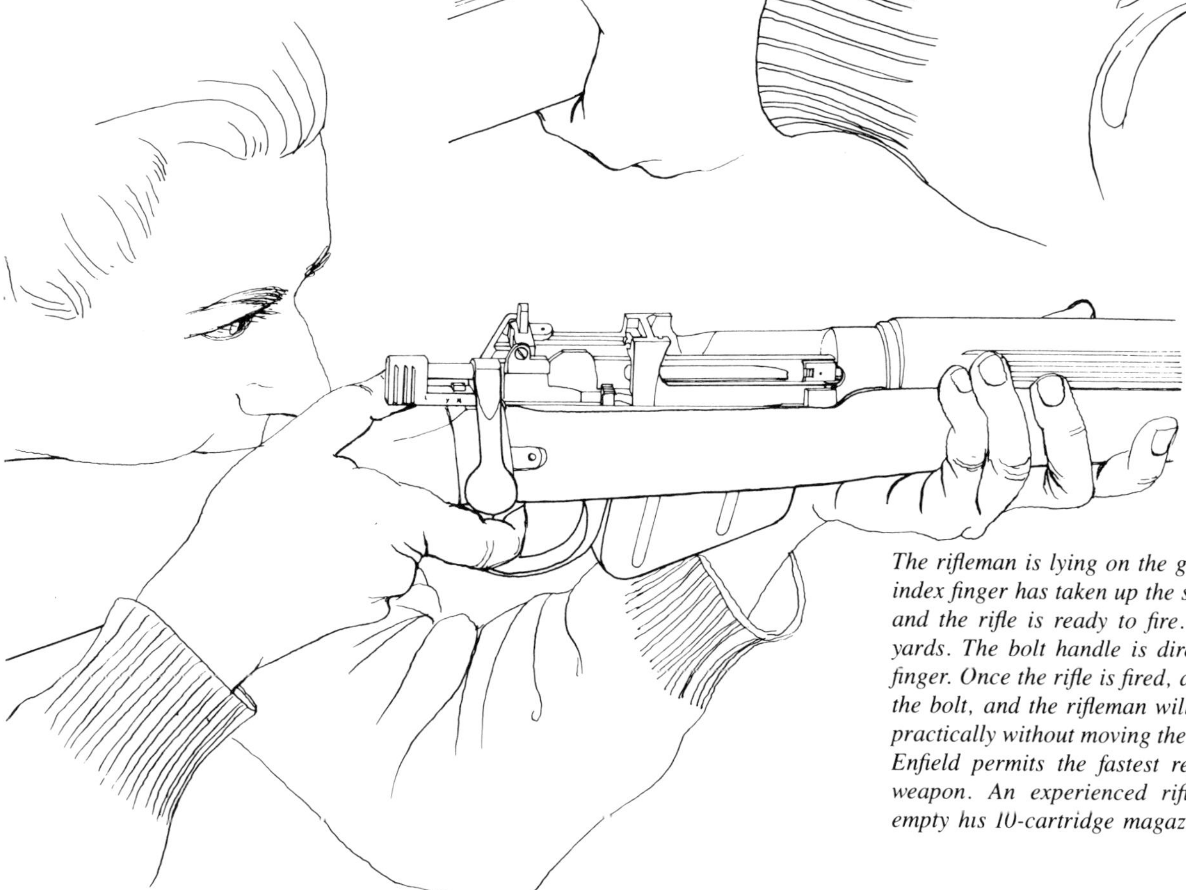

*The rifleman is lying on the ground with his rifle cocked. His index finger has taken up the slack of the double-stage trigger, and the rifle is ready to fire. The battle sight is set for 300 yards. The bolt handle is directly above the rifleman's index finger. Once the rifle is fired, a flick of the index finger unlocks the bolt, and the rifleman will be able to draw back the bolt, practically without moving the rifle from his shoulder. The Lee-Enfield permits the fastest repeating fire of any bolt-action weapon. An experienced rifleman, firing instinctively, can empty his 10-cartridge magazine in 9 seconds.*

We have just reviewed all of the individual arms that were dropped by parachute for the outfitting of saboteur and Maquis units. Even as late as 1943, the deliveries consisted of sabotage items, pistols, and Sten Mark II submachine guns. During the last part of the war, there arrived the first rifles, automatic rifles [called squad light machine guns in Britain] and antitank weapons. Finally, during the last months of the war, a small number of carbines and M-3 submachine guns were dropped.

Submachine guns represented 47 percent of total firearms; rifles, 30 percent; automatic rifles, 5 percent; carbines, 2 percent; antitank weapons, 1 percent; and pistols and revolvers, 15 percent.

Most of the submachine guns were Sten Mark IIs with tubular or skeleton frame stocks. The first of these were delivered along with 4 magazines per gun. The Mark IIIs, with 8 magazines, appeared later. Finally, after D-Day, some well-provisioned companies counted as many as 20 magazines for each gun.

Opinions about the value of the submachine guns varied widely. Some people were wrong in considering the Sten a universal weapon, good for any operation. Others were mistaken in viewing it as a hastily built weapon and a useless toy. The English are fair-minded people who are able to appraise an object at its real value, even if they are its inventors. Consequently, the S.O.E., in our opinion, appraised the Sten very correctly in its sabotage manual:

> The Sten gun is a very deadly close-range weapon but while it can be used effectively at longer ranges, it should not be regarded as a substitute for a rifle. Nothing has been wasted on exterior finish. The barrel and working parts, however, are well made, of good material, and can be relied on to fulfill the purpose of the weapon.
>
> It can be fired at full automatic or at single shot at user's option.

Beginning with the start of American parachute drops in July and August of 1944, the S.O.E. substituted M-3s or Thompsons for the Sten whenever possible. Circumstances, however, made their number minute in comparison to that of the Sten.

The arrival of the Lee-Enfield Mark III* and No. 4 rifles excited the enthusiasm of those who could recognize an extremely fine weapon. It must be admitted that a Lee-Enfield rifle, by its mere appearance, inspired confidence and respect. Delivered with bayonets, they proved to the numerous Resistance fighters that they were now treated as soldiers instead of as bandits. The submachine gun retained, as late as 1942, the taint of American Prohibition gangsters, which made a correct appraisal of its military value impossible for some conservative minds. With a Lee-Enfield rifle, the resistance fighter knew that, for the first time, he was capable of returning fire against Mauser rifles.

But the arm that got the unanimous vote was the M-1 carbine, which, unfortunately, appeared very late and in a very small quantity (less than 10,000). It stirred up a lot of jealousy because it was probably the best gun around for guerrilla operations.

A weapon that had to render the greatest service in actions when long-range, powerful, accurate, and rapid fire was necessary was the automatic rifle. The British supplied what they had: the Bren. Here again, the weapon was exceptional. Fortunately for logistics, the Bren used rifle ammunition.

We will describe this weapon on the following pages.

# AUTOMATIC RIFLE

## Bren Mark I (Bren-Gun) 1937 Model

***Caliber: .303***

*Scale: 1/3*

0 5 10 15 20 25 cm

NOTE: *The Bren gun was a Czechoslovakian 7.92-millimeter weapon adapted to the British .303 cartridge and manufactured under license at the Lee-Enfield arsenal. It was one of the best automatic rifles in the world. Its only disadvantage was that the magazine was above the receiver. This made it necessary to shift the rear sight to the weapon's left side, which made it practically impossible for a left-handed rifleman to use this weapon.*

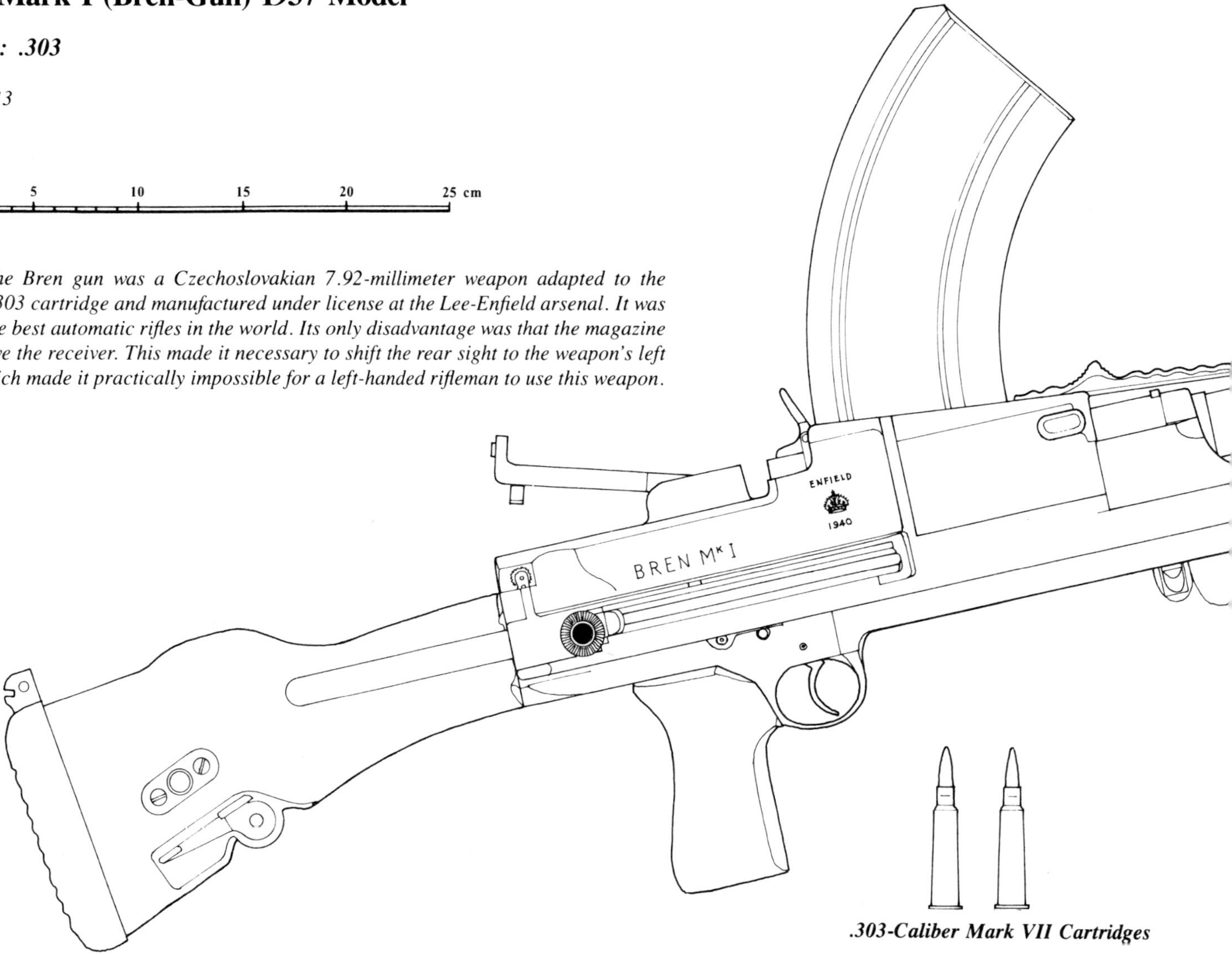


***.303-Caliber Mark VII Cartridges***

*Scale: 1/3*

NOTE: *The weight of the body and the barrel, the sturdy bipod stand, and the length of the sight radius permitted accurate single-shot fire up to 800 yards.*

*The cone-shaped flash hider is for effective night firing.*

*A rotatable grip permitted the weapon to be easily carried with one hand.*



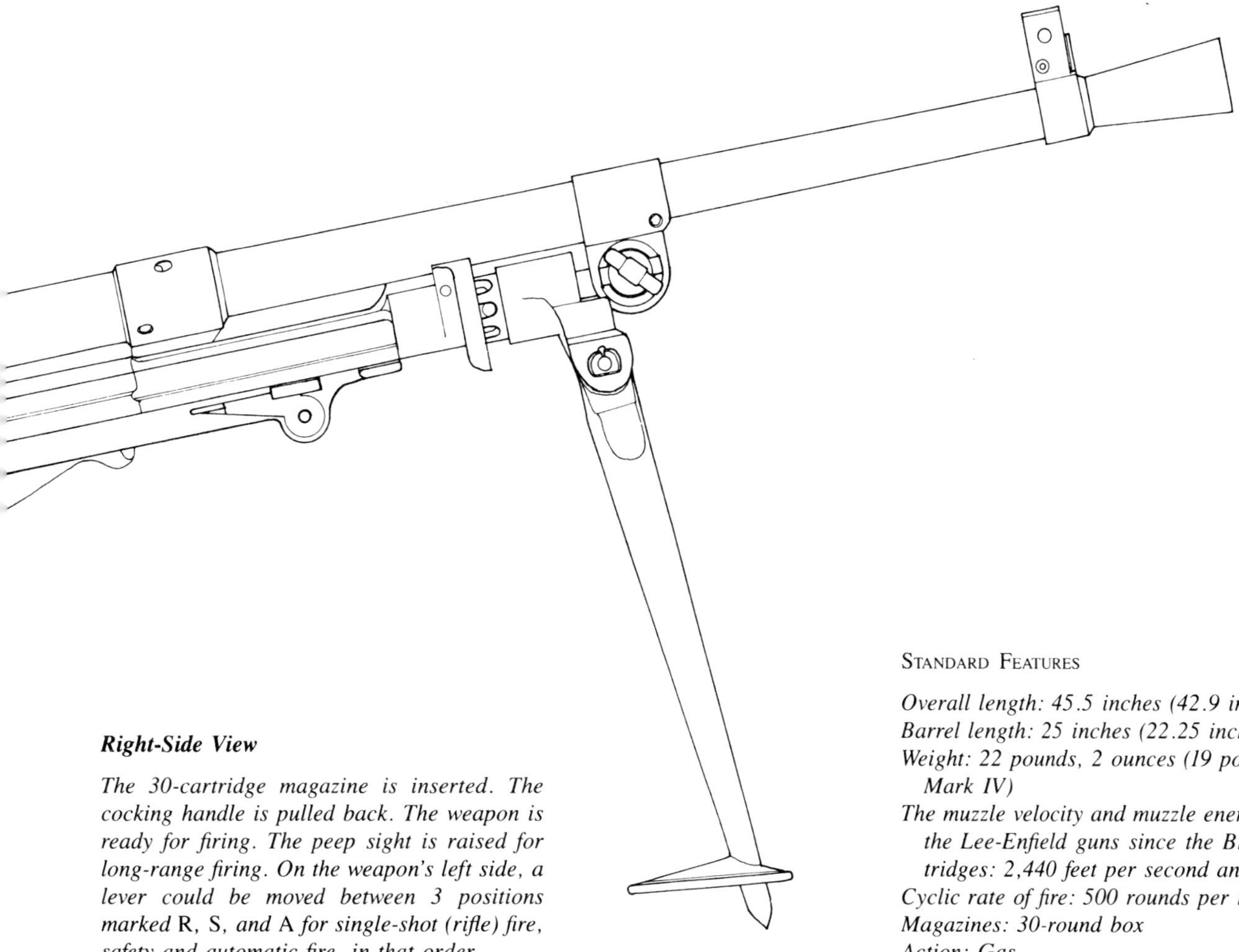

**Right-Side View**

*The 30-cartridge magazine is inserted. The cocking handle is pulled back. The weapon is ready for firing. The peep sight is raised for long-range firing. On the weapon's left side, a lever could be moved between 3 positions marked* R, S, *and* A *for single-shot (rifle) fire, safety and automatic fire, in that order.*

STANDARD FEATURES

*Overall length: 45.5 inches (42.9 inches for the Mark IV)*
*Barrel length: 25 inches (22.25 inches for the Mark IV)*
*Weight: 22 pounds, 2 ounces (19 pounds, 2 ounces for the Mark IV)*
*The muzzle velocity and muzzle energy are the same as for the Lee-Enfield guns since the Bren fires the same cartridges: 2,440 feet per second and 2,310 foot-pounds*
*Cyclic rate of fire: 500 rounds per minute*
*Magazines: 30-round box*
*Action: Gas*

# ANTITANK WEAPONS

## Boyes Rifle

***Caliber: .55***

***Three-Quarter Front View***

***Rifle Ready for Firing***

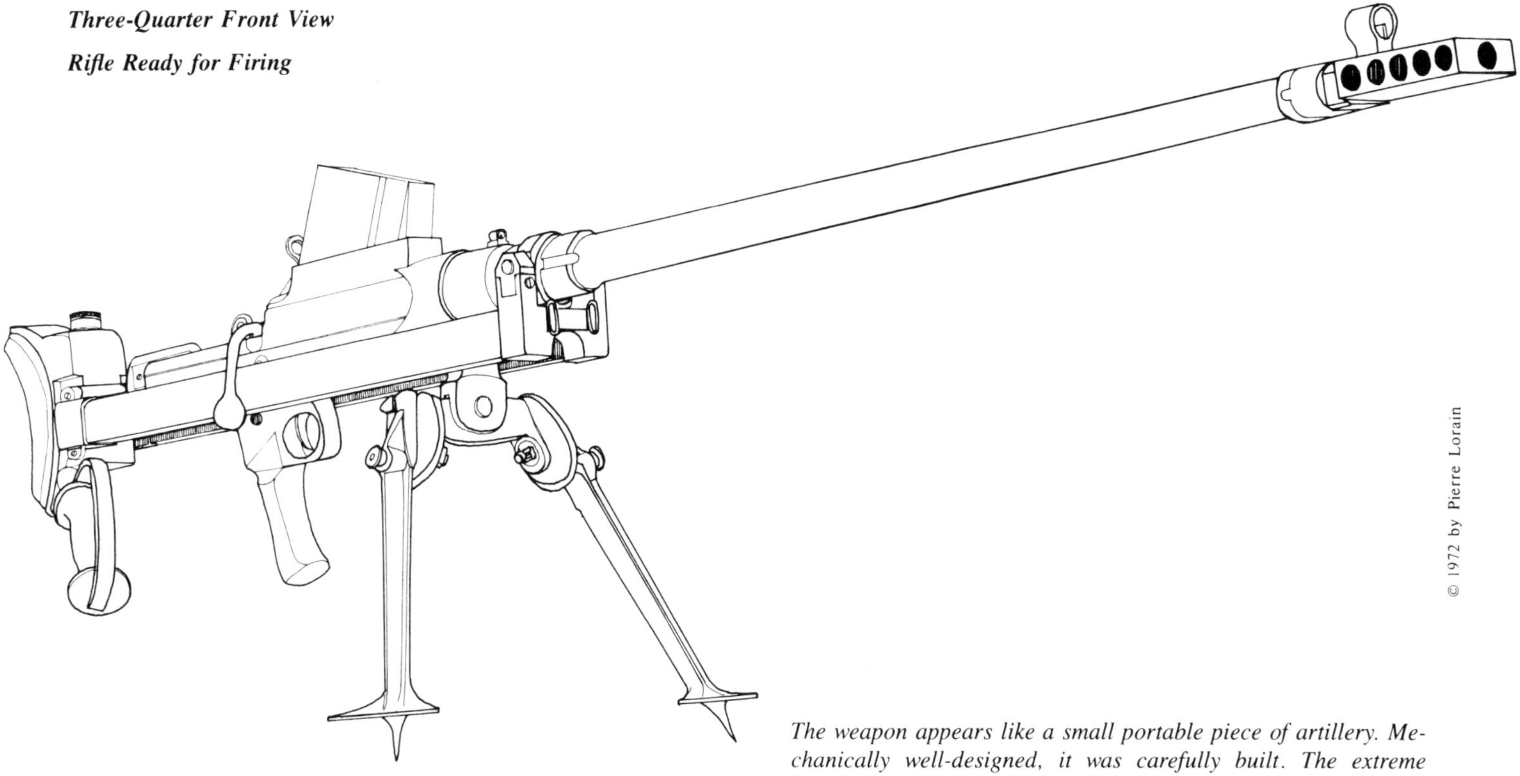



*The weapon appears like a small portable piece of artillery. Mechanically well-designed, it was carefully built. The extreme length of the barrel, which ended in an impressive muzzle brake, the weight of the weapon, and a certain avant-garde modernism give it a singularly efficient appearance.*

*This appearance is deceiving, however. The English had great hopes for this weapon, which was first distributed in 1938, but it proved to be a total failure. It was so heavy that it was practically untransportable. The violence of its detonation and of its recoil made it unpopular from the start. Its ineffectiveness against even the most lightly armored vehicles quickly led to its abandonment. Boyes rifle projectiles could penetrate three-quarters of an inch of steel at 100 yards and half an inch at 500 yards. The steel-core projectiles were nothing more than punches. Its effectiveness against 1939 armored vehicles was almost nil. This mechanism, a direct descendant of 19th-century wall pieces, represented, as did its ancestors, the epitome of a weapon unfit for any military use, especially special operations. However, we describe it here because a few were dropped by the S.O.E. (61, according to Colonel Buckmaster).*

***Details of Boyes Rifle***

***.55 belted rimless cartridges***

*Scale: 1/3*

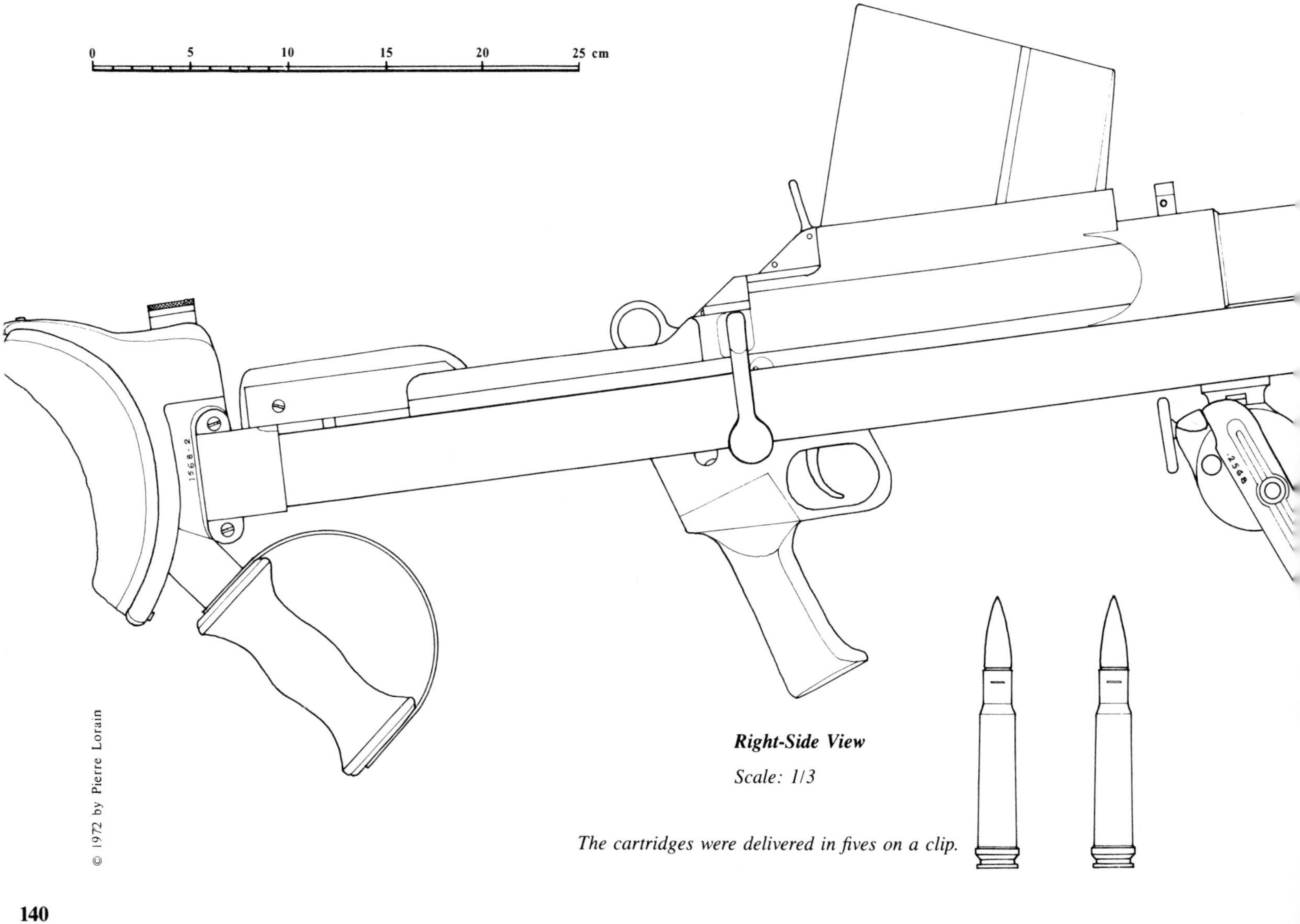


***Right-Side View***

*Scale: 1/3*

*The cartridges were delivered in fives on a clip.*

Note: *Documents give an account of a feat of arms carried out by two soldiers from the No. 4 Commando Unit under the orders of Lord Lovat, using a Boyes rifle. This action took place at dawn during the raid on Dieppe on August 19, 1942. A few point-blank shots of 14-millimeter projectiles silenced a German 20-millimeter automatic cannon that was mounted on an anti-aircraft tower near Varengeville and that had stopped the advance of the Commandos.*

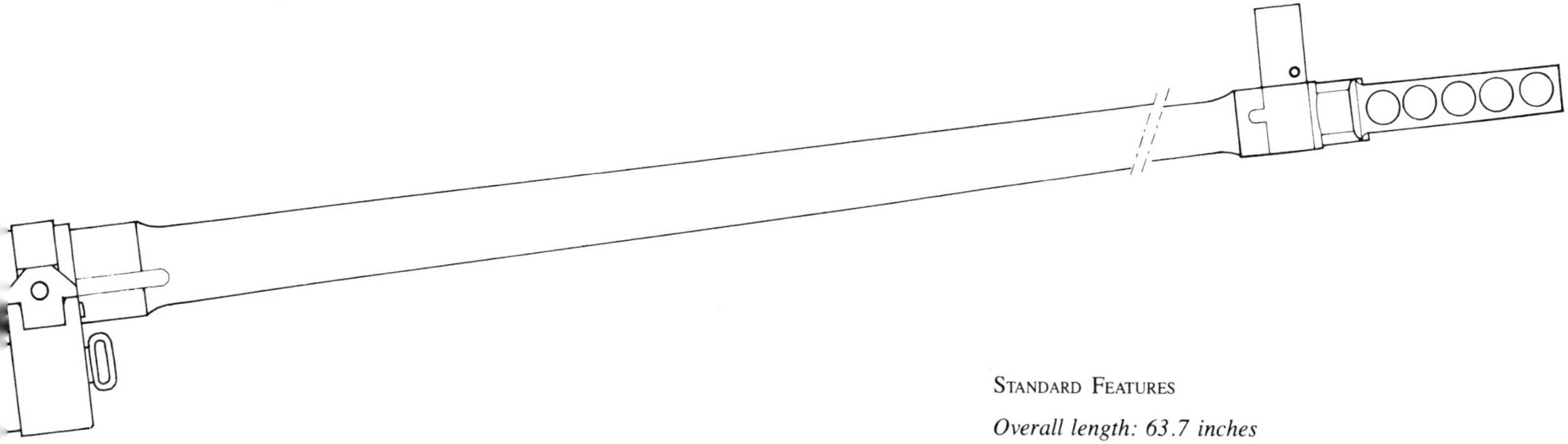

Standard Features

*Overall length: 63.7 inches*
*Barrel length: 36 inches*
*Weight: 36 pounds 8 ounces*
*Muzzle velocity: 26,000 feet per second*
*Projectile weight: 3 ounces*
*Muzzle energy: 12,000 foot-pounds*
*Rate of fire: 6 to 8 shots per minute*
*Bolt action weapon*
*Magazine capacity: 5 armor-piercing cartridges or 5 tracer cartridges*

*The 5-cartridge magazine is inserted. The breech is closed and locked. The hammer is uncocked. The weapon is operated like a repeating rifle. After each shot, the gunner must recock and unlock and pull back the bolt. This pulls out the spent cartridge, which falls to the ground under the gun carriage. Moving the bolt forward slides the first cartridge from the magazine into the chamber. Once the bolt is pulled back down, the weapon is again ready for firing.*

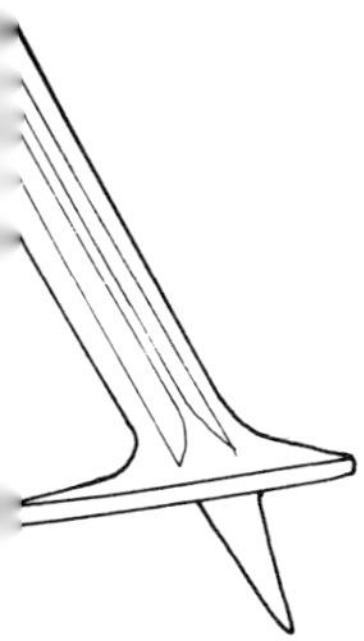

*The recoil is substantial: a rubber cushion is attached to the butt plate. When the shot was fired, the weapon recoiled against its carriage like a piece of artillery. The muzzle brake was effective, but rendered the detonation unbearable.*

*The weapon had a simplified double-marked rear sight for firing at 300 and 500 yards. The magazine's location on top of the receiver and in the firing axis required that the sight line be moved to the left side. Accordingly, only a right-handed gunner could use the weapon.*

# Rocket Projector (Officially Called "Projector, Infantry, Antitank" or "P.I.A.T.")

## *Projectile: Hollow-Charge Rocket*

*ADOPTED JUNE 5, 1943*

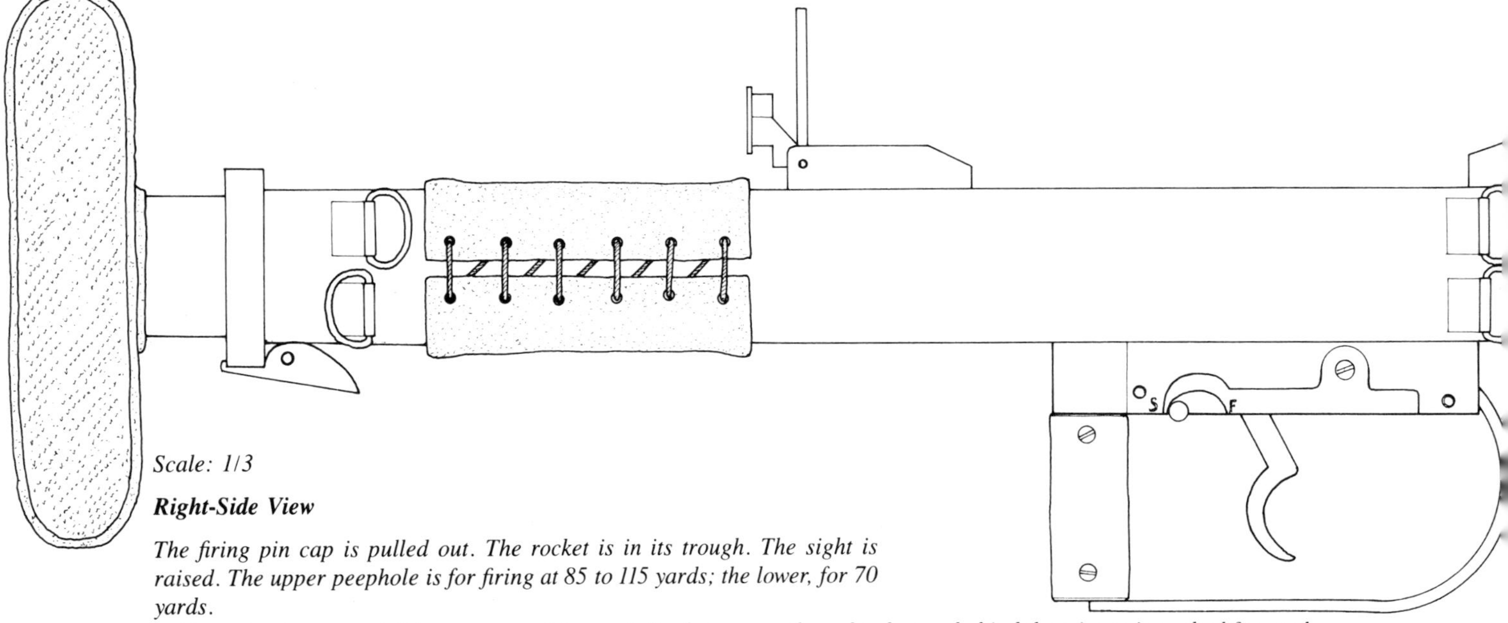


*Scale: 1/3*

***Right-Side View***

*The firing pin cap is pulled out. The rocket is in its trough. The sight is raised. The upper peephole is for firing at 85 to 115 yards; the lower, for 70 yards.*

*The safety is engaged. The weapon will be ready for firing once the safety button behind the trigger is pushed forward to* F. *The 4 rings welded onto the body of the weapon hold the 2 standard webbed straps for carrying.*

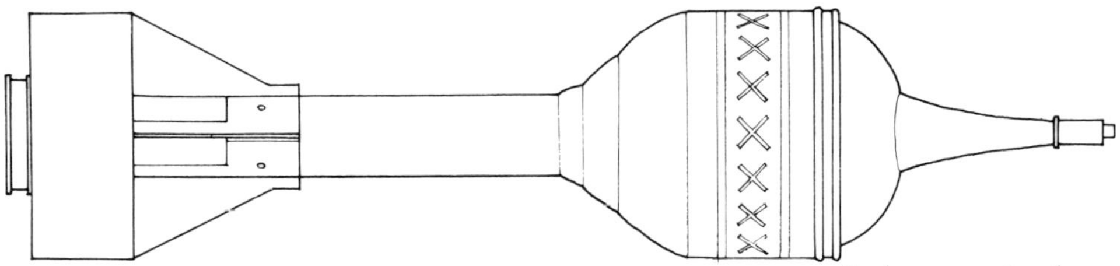

***Hollow-Charge Rocket Type A.T. M-1 A-1***

*Scale: 1/3*

*Rockets are painted green with a band of red* X*s around the charge. As a safety measure, the percussion fuse is not attached to the rocket until the last minute. During transport, the fuse is placed in a watertight package, which is attached to the tail of the rocket with a spring clip.*

Standard Features

*Overall length: 39 inches*
*Weapon weight: 34 pounds, 5 ounces*
*Rocket weight: 2.5 pounds*
*Maximum usable range (antitank): 115 yards*
*Maximum usable range (mortar): 350 yards*

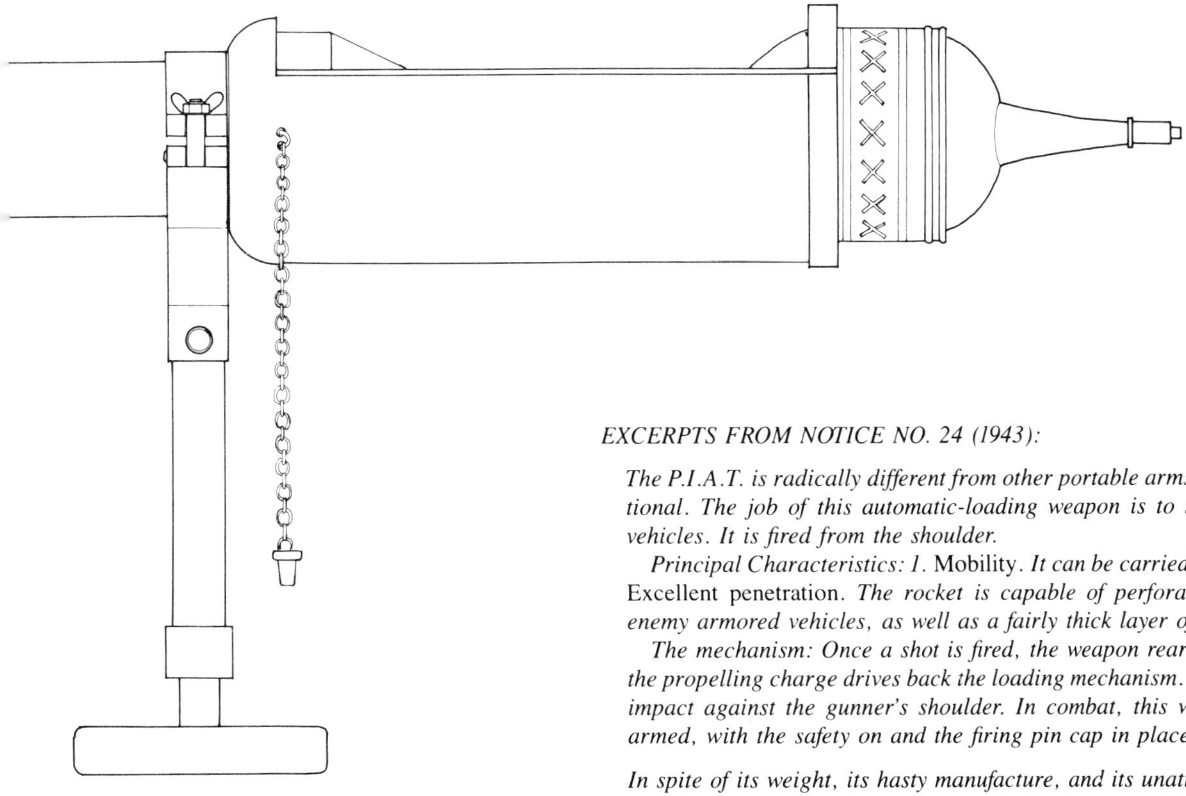



*EXCERPTS FROM NOTICE NO. 24 (1943):*

> *The P.I.A.T. is radically different from other portable arms. It is as simple as it is unconventional. The job of this automatic-loading weapon is to stop and destroy enemy armored vehicles. It is fired from the shoulder.*
>
> *Principal Characteristics: 1.* Mobility. *It can be carried by one soldier. 2.* Short range. *3.* Excellent penetration. *The rocket is capable of perforating the armor of all late-model enemy armored vehicles, as well as a fairly thick layer of reinforced concrete.*
>
> *The mechanism: Once a shot is fired, the weapon rearms itself automatically. Gas from the propelling charge drives back the loading mechanism. A strong spring reduces the recoil impact against the gunner's shoulder. In combat, this weapon should always be carried armed, with the safety on and the firing pin cap in place.*

*In spite of its weight, its hasty manufacture, and its unattractive appearance, the P.I.A.T., unlike the Boyes antitank gun, was probably the most effective weapon of its time. Adopted June 5, 1943, it proved to be one of the principal pieces of the clandestine arsenal and often showed its effectiveness during attacks against rail or road convoys and fuel depots.*

*A total of 1,205 P.I.A.T.s, with 20 rockets each, were parachuted to the French Resistance.*

*The inclination of the tube in the illustration corresponds to the angle of projection for firing at 100 yards.*

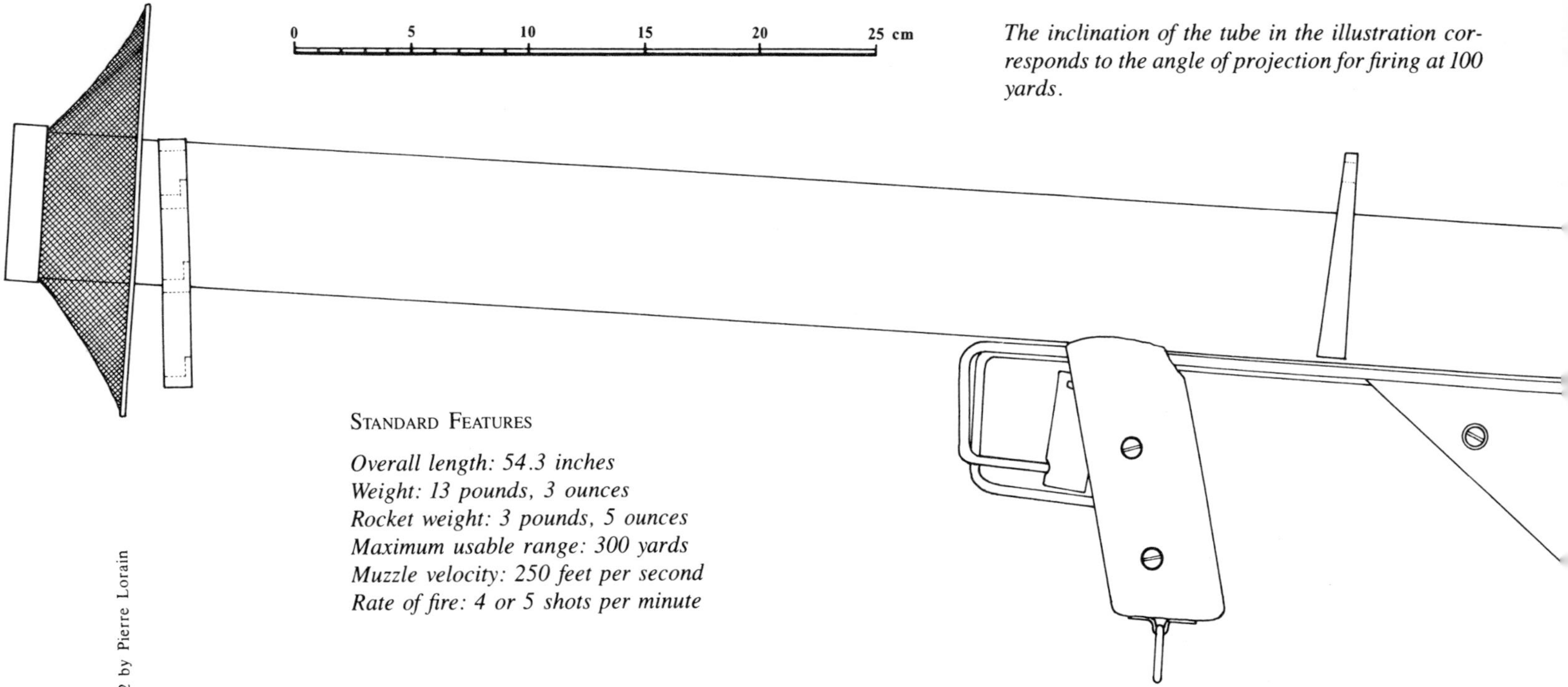


Standard Features

*Overall length: 54.3 inches*
*Weight: 13 pounds, 3 ounces*
*Rocket weight: 3 pounds, 5 ounces*
*Maximum usable range: 300 yards*
*Muzzle velocity: 250 feet per second*
*Rate of fire: 4 or 5 shots per minute*



## *Type A.T. M-6 A-1 60-Millimeter Rocket*

*Scale: 1/3*

*The rockets are painted olive green with identification marks in yellow. The propelling charge consists of 5 ballistite tubes, which are ignited by an electrical primer. The inertial percussion device is permanently fixed in the rocket's body and is protected by a pin which is pulled out just before firing. The hollow bazooka charge is less effective than that of the P.I.A.T., but it could penetrate and destroy the heaviest enemy tanks.*

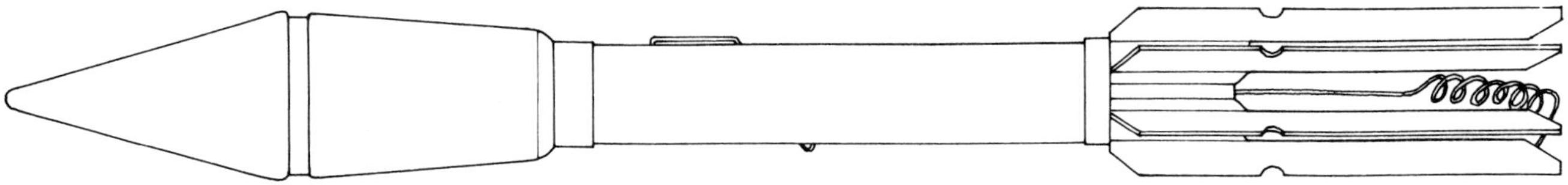

# Rocket Launcher Model A.T. M-1 A-1 (Bazooka)

***Projectile: Hollow-Charge Rocket. Caliber: 60-Millimeter***

*Scale: 1/3*

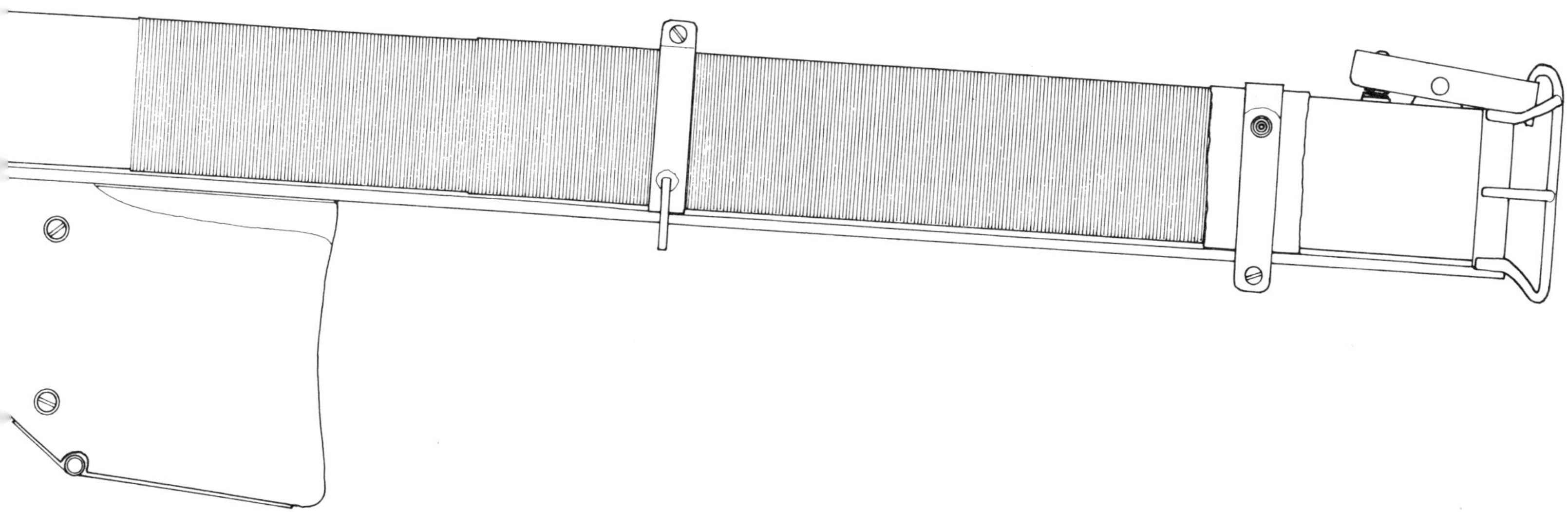

*The bazooka is made up of a simple tube of thin steel sheet, open at both ends and serving as a launching ramp for the rocket. Iron wires are bound around part of its length. An electrical circuit fed by 2 dry-cell batteries sets off the firing mechanism when the gunner pulls the trigger. This weapon has no other mechanism.*

*On the front half of the weapon are the fire deflector grid, the 3 front sights for 100, 200, and 300 yards, a pistol grip with electric trigger, the peepsight, a wooden shoulder butt containing the dry-cell batteries and a flashlight bulb to check whether the current is adequate, and finally, a ring serving as a lower band. At the back of the weapon, there is a second ring insulated from the body of the bazooka by a plastic band; on this ring is mounted an electrical spring contact. Finally, there is a lock which closes over the rocket's fins to prevent it from slipping out.*

*The bazooka requires a 2-man team to operate it. The loader inserts the rocket through the back until the lock clenches onto one of the projectile's fins. He then pulls the electrical wire out from the rocket and wraps the bare end around the spring contact. He jumps away from the weapon and slaps the gunner on the shoulder to let him know it is loaded. When the gunner fires, a burst of flame shoots from the back of the tube.*

# Principle of the Hollow Charge (P.I.A.T. and Bazooka)

*1. CROSS SECTION OF THE SHELL*

*From left to right, there are the inertial firing pin and its spring, the primer and detonator, the shell filled with the explosive charge and hollowed out in the form of a crater. The steel-sheet pointed-arch housing serves only to give an aerodynamic shape to the projectile and to place the charge, at the moment of impact, at the optimal distance from its objective.*

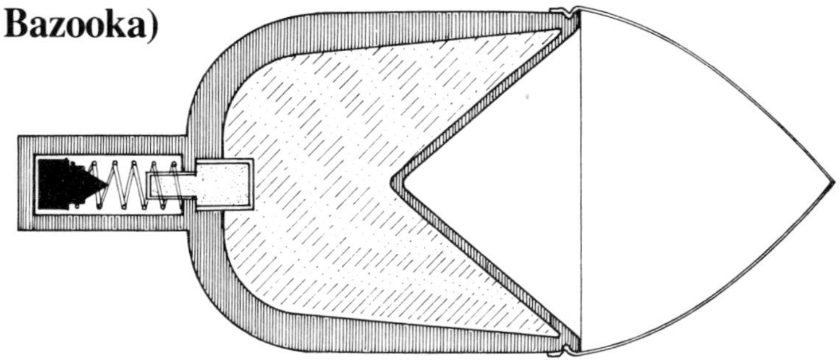

*The velocity of the projectile has nothing to do with its destructive power. The propellant charge consists of a rocket which easily carries the hollow charge to its target.*

*2. MOMENT OF IMPACT*

*The nose strikes the armor plate and starts to collapse. The firing pin is thrown forward and strikes the primer, which sets off the charge. The shock wave follows the lines of least resistance and focuses along the axis of the hollow cone.*

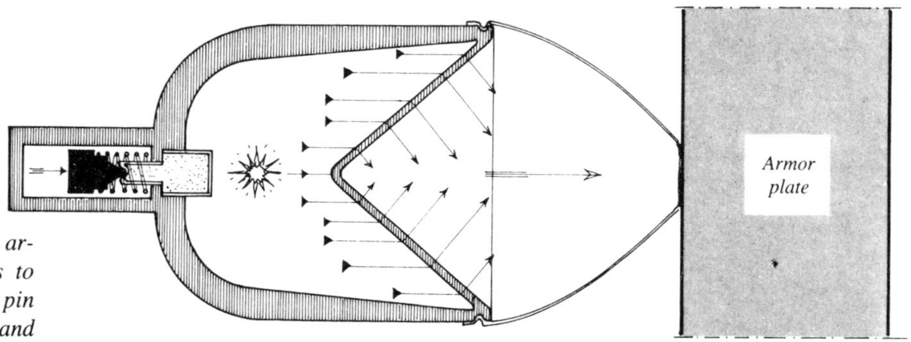


*3. SHELL DISINTEGRATION*

*Almost all the force is concentrated in an ultra-high-pressure gas jet whose speed has reached a velocity of 30,000 feet per second by the time it strikes the armor. The armor melts instantly. Gas velocity is still 12,000 feet per second. Following the gas jet, secondary projectiles composed of melted armor and of a core formed by parts of the fused cavity casing disperse at great speed inside the tank, which immediately catches fire.*

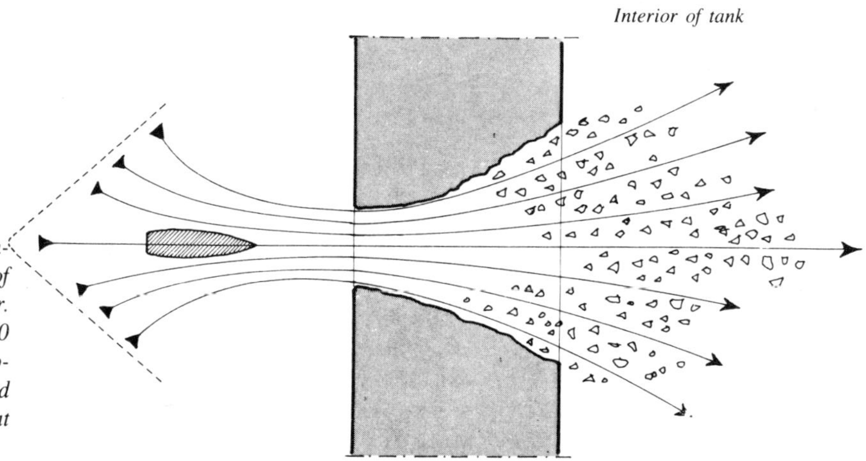

# SILENCER WEAPONS OF THE SECRET SERVICE

# Principles of Silencers

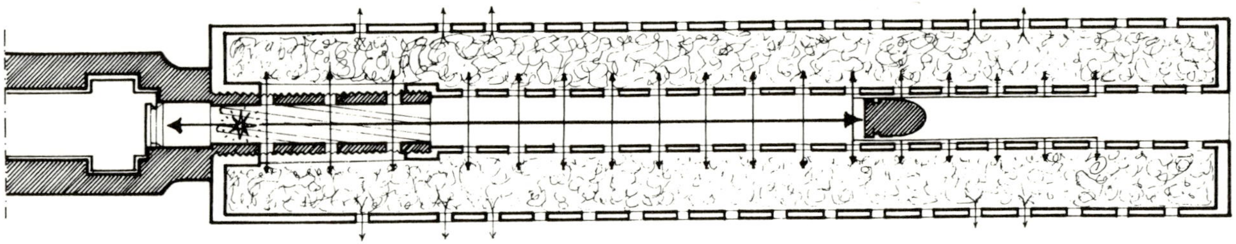

*First Principle*

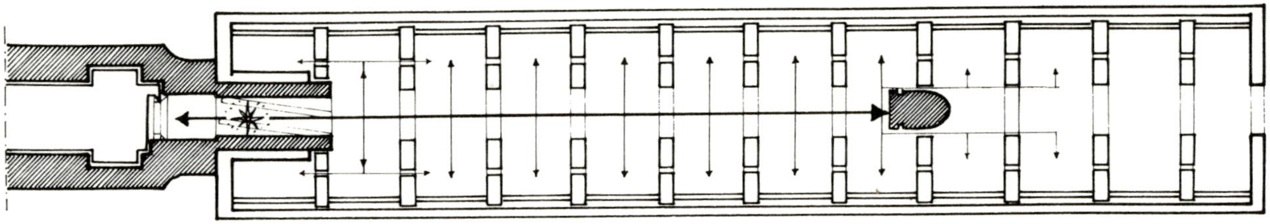

*Second Principle*



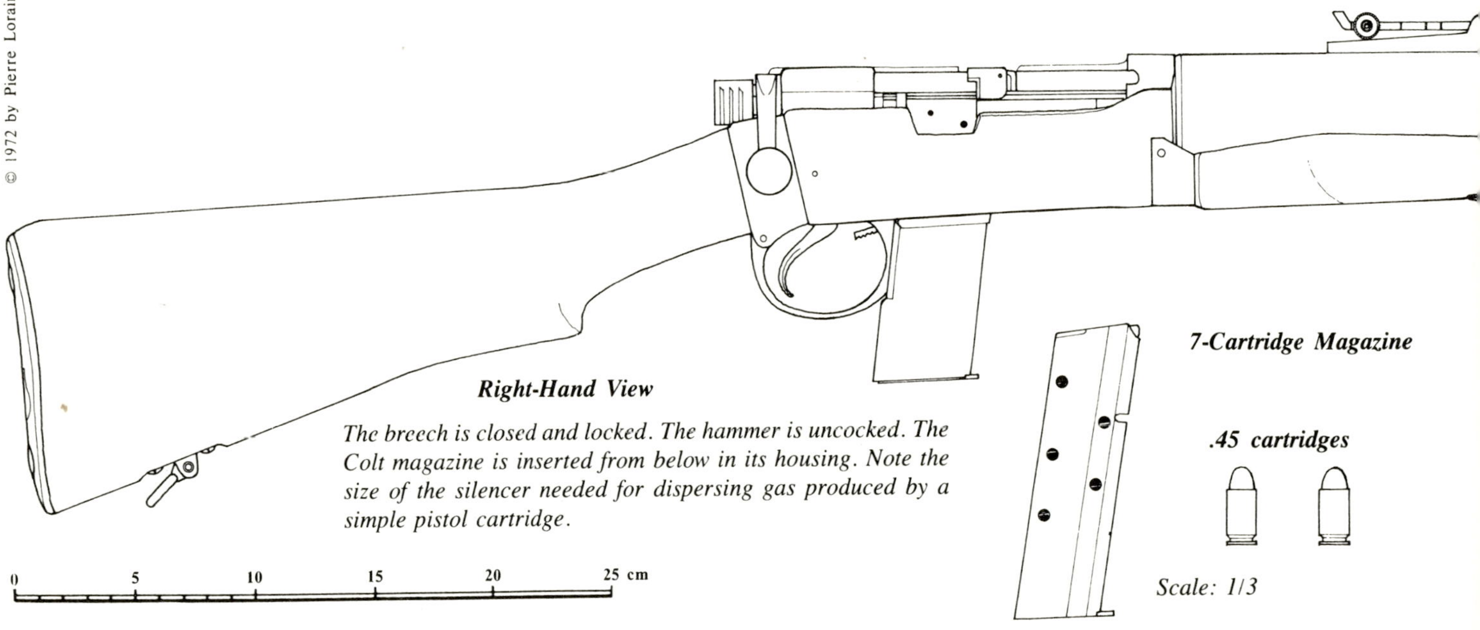

***Right-Hand View***

*The breech is closed and locked. The hammer is uncocked. The Colt magazine is inserted from below in its housing. Note the size of the silencer needed for dispersing gas produced by a simple pistol cartridge.*

***7-Cartridge Magazine***

***.45 cartridges***

0 5 10 15 20 25 cm

*Scale: 1/3*

*The sudden and instantaneous expansion of gas at the muzzle produces a detonation.*

*Therefore, the function of a silencer is to slowly and progressively reduce the pressure of the combustion gas, until the noise of the detonation is almost completely nullified.*

*FIRST PRINCIPLE*

*The body of the silencer is made up of a perforated sheet-steel center tube with a diameter slightly greater than the projectile's caliber (about 1/16 inch more) and of an outer casing, also a perforated steel sheet.*

*The space between the two casings is filled with steel wool. The entire device is solidly attached to the barrel, which may itself be perforated.*

*When the shot is fired, the gas slowly diffuses through the steel wool. If the silencer is long enough, there is no audible detonation when the projectile passes the mouth.*

*SECOND PRINCIPLE*

*The body of the silencer is made up of a cylindrical sleeve which is completely enclosed, except at its ends. Inside this sleeve, round disks of light alloy, pierced by holes, are held spaced apart by rings or springs and form expansion chambers and baffle plates for diffusing the gas as the projectile passes through. The effectiveness of the system is proportional to the number of expansion chambers and their volume.*

NOTE: *When a silencer is used, ammunition offering a muzzle velocity substantially less than the speed of sound is required, for any supersonic projectile produces a violent crack as it passes through the air.*

*Usable are 7.65-millimeter and Colt .45 cartridges. However, the Parabellum 9-millimeter cartridge must have a specially reduced charge. The standard long rifle .22 cartridge is a borderline case: it can be noisy during the trajectory of its projectile.*

## De Lisle Silent Commando Carbine Mark I

***Caliber: .45***

*Scale: 1/3*

*Named after its inventor, this carbine was completely silent. It was designed for special commando operations and was widely employed by the many British secret services, especially for the defense of Intelligence Service bases or clandestine radio stations. It was modeled after a Lee-Enfield Mark III rifle, but the breech and the magazine were modified to accommodate a .45-caliber 7-cartridge Colt pistol magazine.*

*A new, very short, perforated barrel of .45 caliber was added. It was surrounded and lengthened by a patented integrated silencer. The weapon was extraordinarily effective: the only audible sounds were the maneuvering of the bolt and the hammer striking the cartridge primer. The recoil was insignificant, and the accuracy excellent up to 150 yards.*

# "Welrod" Silent Pistol

***Caliber: .32***

*Scale: 1/3*

NOTE: *This weapon did not carry any identification mark. It could have been either British or American in origin. It was freely given to S.O.E. and I.S. agents.*



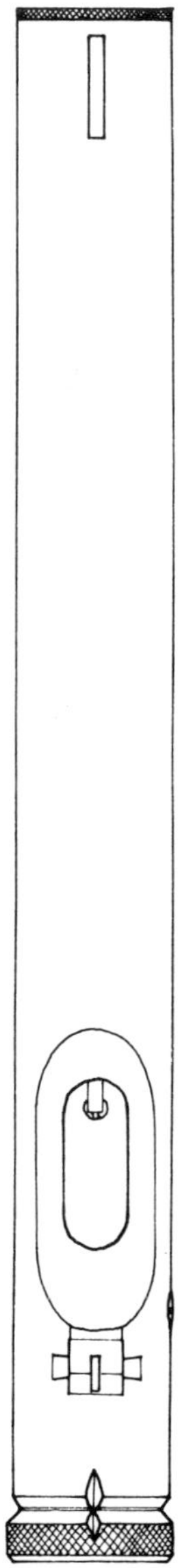

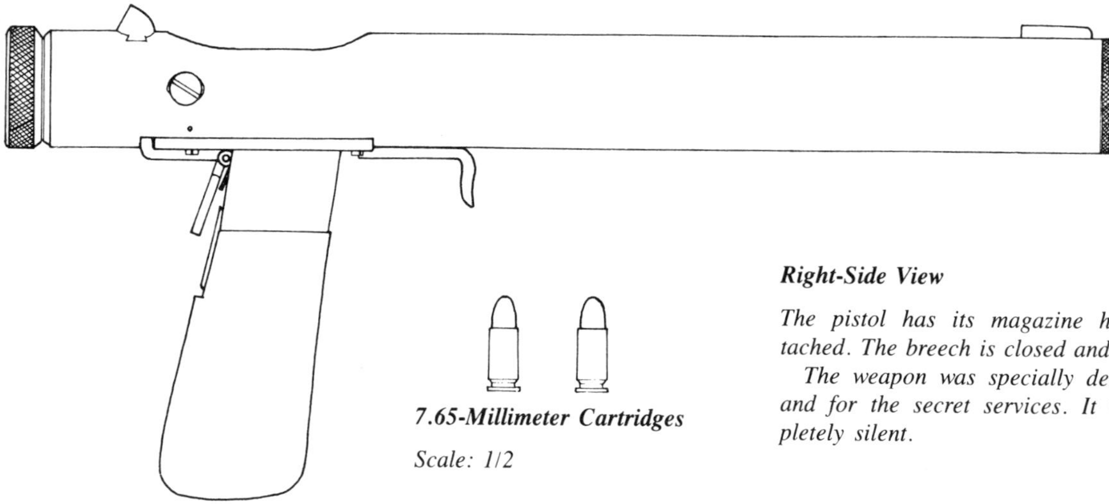

***7.65-Millimeter Cartridges***

*Scale: 1/2*

***Right-Side View***

*The pistol has its magazine handle attached. The breech is closed and locked.*

*The weapon was specially designed by and for the secret services. It was completely silent.*

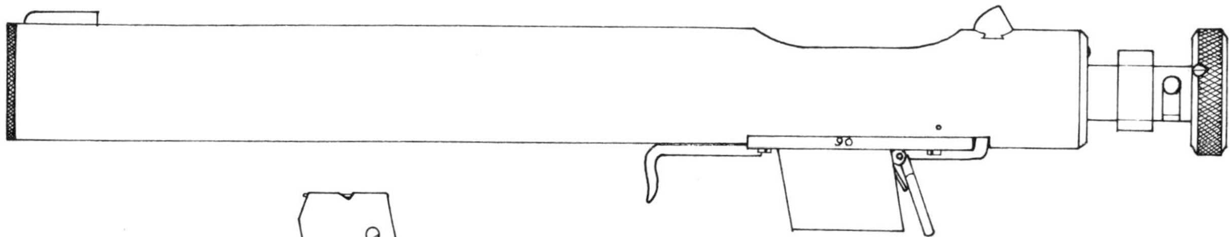


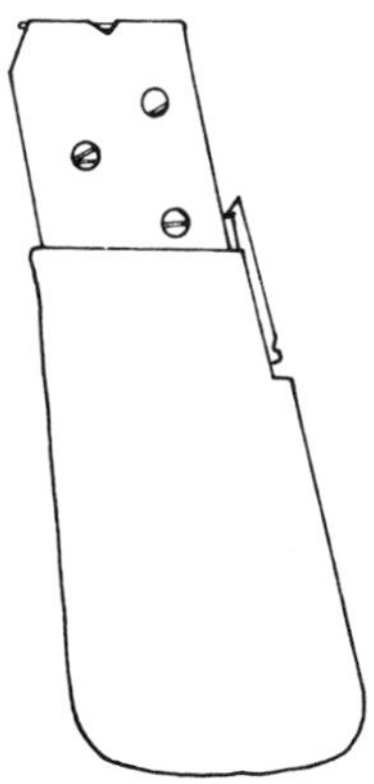

*The silencer is an integral part of the barrel.*

***Top View***

*The breech is closed and locked.*

*The shape, dimensions, and volume of the barrel won it the nickname "bicycle-pump."*

***Left-Side View***

*The pistol is separated from its magazine handle.*

*The bolt is open, unlocked, and pulled back.*

*After each shot, it is necessary to move the bolt manually to eject the spent cartridge and introduce a new cartridge from the magazine into the chamber.*

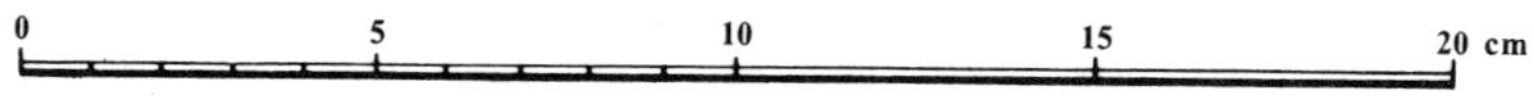

## American High-Standard Pistol (with Silencer)

***Caliber: .22 long rifle***

*Scale: 1/2*

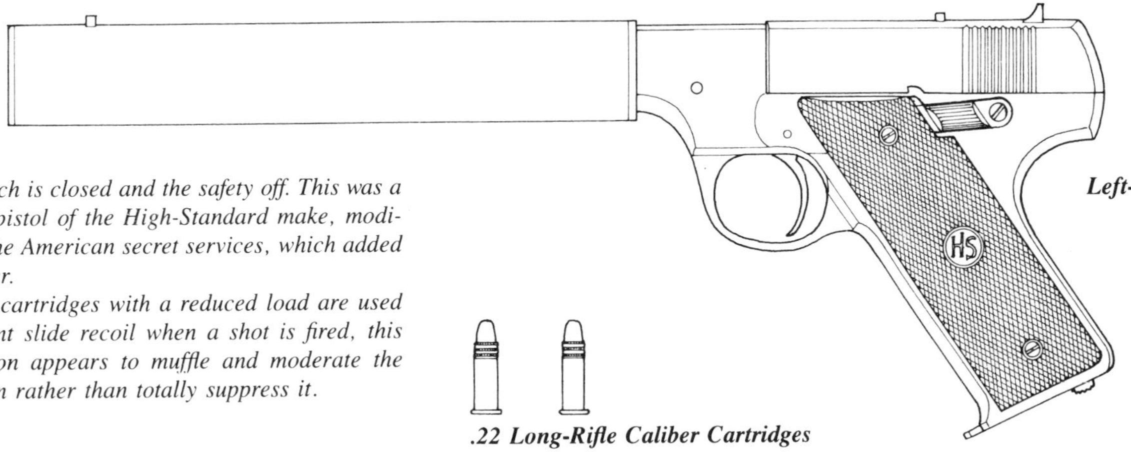

***Left-Side View***

*The breech is closed and the safety off. This was a civilian pistol of the High-Standard make, modified by the American secret services, which added a silencer.*

*When cartridges with a reduced load are used to prevent slide recoil when a shot is fired, this adaptation appears to muffle and moderate the explosion rather than totally suppress it.*

***.22 Long-Rifle Caliber Cartridges***

*Scale: 1/2*

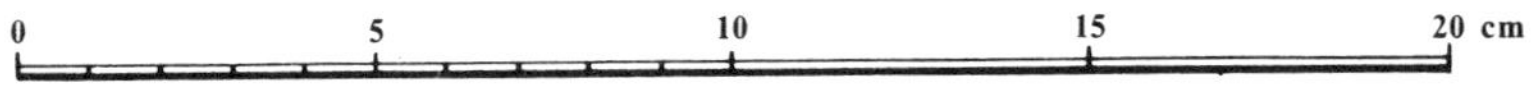


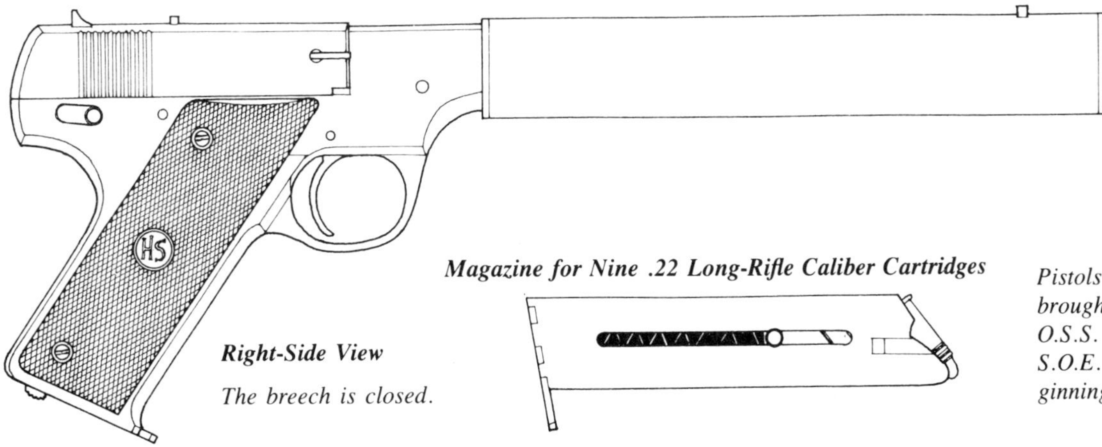

***Right-Side View***

*The breech is closed.*

***Magazine for Nine .22 Long-Rifle Caliber Cartridges***

*Pistols of this type were apparently brought to France with the first O.S.S. agents integrated into the S.O.E. and dropped by parachute beginning in June 1944.*

## Sten Mark II Submachine Gun (with Silencer)

***Caliber: 9-Millimeter Parabellum***

*Scale: 1/3*

0 5 10 15 20 25 cm

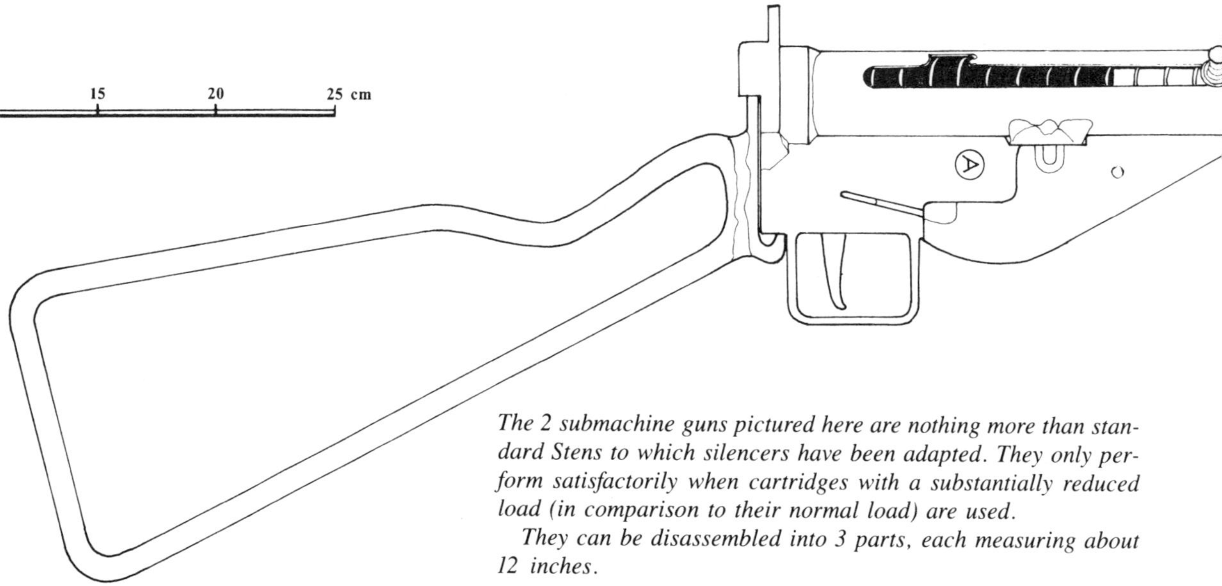

*The 2 submachine guns pictured here are nothing more than standard Stens to which silencers have been adapted. They only perform satisfactorily when cartridges with a substantially reduced load (in comparison to their normal load) are used.*

*They can be disassembled into 3 parts, each measuring about 12 inches.*

## Sten Mark V Submachine Gun (with Silencer)

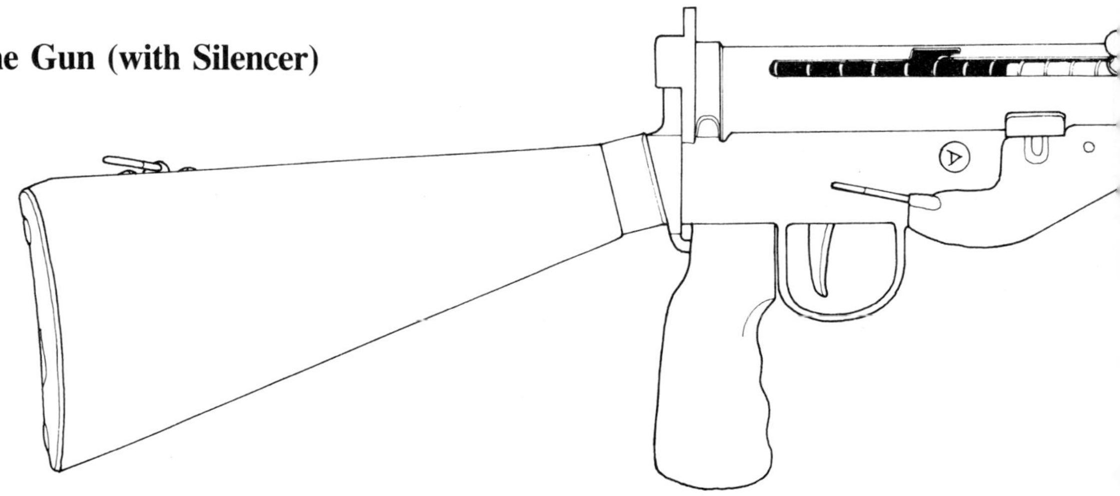

*Since the silencer casing is flush with the receiver, the grip attached to the radiator on the original Mark V had to be omitted, the shooter's left hand now grasping the silencer. Note the laced webbing sleeve which offers a better grip.*

*The silent Mark II and the Mark V were widely distributed to S.O.E. agents who were proven marksmen for special operations, especially for defending radio stations in towns. These guns were used with the De Lisle carbine.*

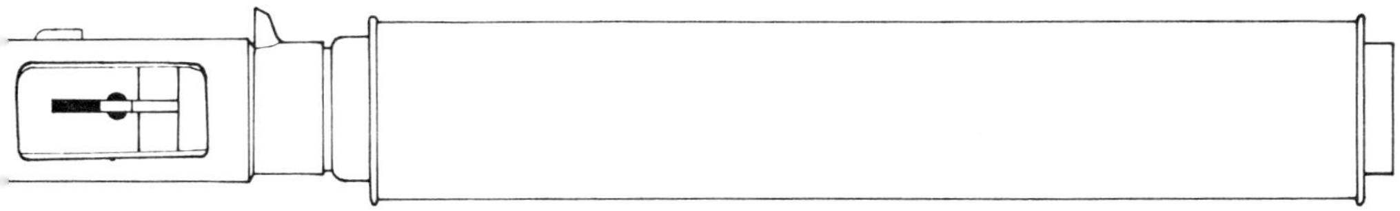

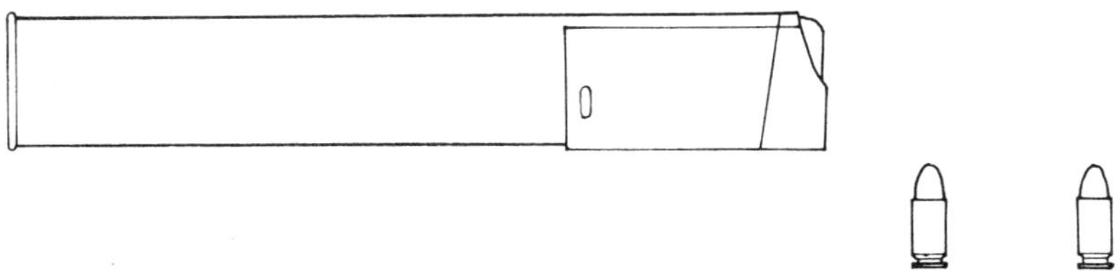

***32-Cartridge Magazine and Parabellum 9-Millimeter Cartridges***

*A special load [charge] called the silent load brought down the muzzle velocity to about 800 feet per second.*

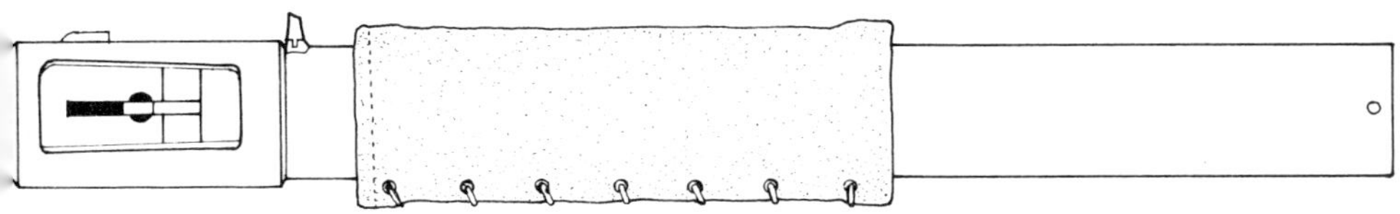

NOTE: *Functioning in a classic way, these silencers housed several rubber disk membranes, pierced in their centers with a small hole just allowing the bullet to pass through. Since heat is dissipated with each shot, the membranes hardened, then burned fairly quickly. After 15 shots or so, the cylinder had to be dismantled and the rubber membranes replaced; otherwise, the effectiveness of the system would have been considerably diminished. If the membranes were new and special cartridges were used, the detonation was almost entirely suppressed. However, only single shots could be fired. After each shot, the rifleman had to eject the spent cartridge and reload by pulling the bolt back. But the dry click that the breech block produced as it closed up abruptly and struck the cartridge when the rifleman pulled the trigger could not be avoided.*

# SABOTAGE AND EXPLOSIVES

Since sabotage was one of the principal operations of the S.O.E., it was a good idea to teach the rudiments, not so much to the agents, who were already experts in the subject, but to the men they could recruit. To this end, the S.O.E. played publisher and added a small anonymous brochure to each container. In 7 languages, this brochure described how to best use the extraordinary panoply of unusual, diabolical, and deadly objects which it modestly called items.

Even if this work is not meant to be read by everyone, it is occasionally lively. We shall give a few excerpts.

The special sabotage materials distributed by the S.O.E. include explosive and incendiary devices, and booby traps and snares.

## Explosive Devices

As a security measure, the various components of explosive devices were packaged and separated one from the other. They were assembled when the device was to be used.

Accordingly, yellow explosive "plastic" was offered in the form of 4- or 8-ounce sticks, 5-pound slabs, or in ready-to-use charges in special packages, such as those for railroad sabotage, or for general applications in 500-gram (1-pound) boxes or 250-gram (half-pound) "clams."

If no explosive "plastic" was available, Nobel "808" in the form of 4-ounce cartridges, brown in color and smelling strongly like bitter almonds, replaced it. "When handling 808, do not handle for long with bare hands, or you will get a bad headache."

Primers were made in a shape like a cork or delivered in Cordtex rolls for connecting the charges one to the other.

Detonators were delivered in antishock packages stuffed with sawdust.

There were 3 kinds of igniters:

Instant automatic: "fog signal" and all traps

Delayed automatic: pencil time fuse (acid bulb or L type)

Delayed-action manual: Bickford safety fuse (lit with a special nonflaming match called a fusée)

The most liked ready-to-use charges were the limpet and clam magnetic mines. The first ones, which we will describe in detail later, were meant for the sabotage of ships. Six magnets stuck the mine firmly to the hull. The explosion resulted in a breach that was impossible to patch. The color of the acid bulb determined the delay. Assuming the water temperature to be 40° Fahrenheit, a red bulb would give a delay of 4½ hours, and a violet bulb, 5½ days. Orange, yellow, green, and blue provided delays between those of red and violet.

These Mark I Limpet mines were used during the Frankton operation of December 7, 1942. A 5-canoe naval Commando unit went up the Gironde estuary and seriously damaged 4 docked ships.

The Mark III Limpet mine, which appeared during the last part of the war, was shaped like a dinner plate. It used a standard L-shape delayed-action pencil time fuse instead of the fragile acid bulbs.

A magnetic pocket-size mine called the clam was made up of a small bakelite box filled with 8 ounces of explosives. It was equipped with 2 magnets and was quite often used to blow up engine parts, locomotive cylinders, and automobiles.

> For locomotives: always place the charge on the right-hand side cylinder. This will prevent the Germans using an undamaged cylinder from another locomotive to replace the damaged one.

Railroad charges were used with the "fog signal" type igniter pictured herein.

The tyre burster device, shaped like a face-powder compact, was a small steel mine which exploded under 100 pounds of pressure. It could destroy the heaviest of tires. The S.O.E. suggested other uses for these mines:

> b: The "tire puncture" [tyre burster] might also be inserted: . . . c: on the driver's seat of a car, placed under the covering; d: under a carpet or mat . . .

Detailed illustrations of the Gammon grenade are given. This was one of the most efficient S.O.E. weapons.

Before throwing this formidable device, it was a good idea to open one's mouth and fill it with cotton wool to avoid popping one's eardrums. "The Gammon is most useful against armored cars, staff cars and trucks. It can also be used against light tanks, but should not be used against heavy tanks."

To increase its antipersonnel effect, it was customary to insert nails or shell fragments in the bulb of the explosive—as long as one could take shelter upon throwing the grenade.

## Incendiary Devices

The S.O.E. supplied incendiary cigars, a Mark I "Firepot" filled with Thermit and a 1-kilogram (2¼ pound) Thermit bomb:

> The bomb burns quickly, generating intense local heat and fierce flames. It will penetrate mild steel plate 3.1 mm thick.
>
> Its main uses are:
>
> (i) For igniting petrol and other inflammable oils in closed containers, e.g., army petrol cans in stacks.
>
> (ii) For attacking difficult non-inflammable targets of compressed material, in conjunction with other incendiaries.
>
> (iii) For demolition work, e.g., for attacking telecommunication cable, electric motors, etc.

The S.O.E. also supplied an oxygen-producing "incendiary block." It was particularly effective for setting parcels in packing cases on fire through their lids, as well as in poorly ventilated warehouses.

Finally, the S.O.E. also supplied the pocket time incendiary. After removing the tiny colored wooden rod from its slot, the acid bulb was crushed with the edge of a coin. This small device burned fiercely for about 50 seconds. The rod's color indicated the delayed reaction. The color code was the same as that of the pencil time fuses: red, ½ hour; white, 2 hours; green, 5½ hours; yellow, 12 hours; blue, 24 hours.

## Tricks

We will classify under "tricks" all instant-pressure, depressure, and draw-cord igniters which imitated any familiar object. Specifically, there were explosive gravel, which was used on roads; explosive anthracite pieces, which were mixed with coal stockpiles feeding ports and railway stations; and the "stuffed rat," or rat skins stuffed with explosive plastic and equipped with a detonator and a small piece of Bickford fuse, to be planted in ship stokeholds. The stoker's natural reaction would be to pick the rat up with his shovel and throw it into the boiler, causing it to explode.

We should also mention the abrasive paste used to grease axle boxes of railroad rolling stock and the altimetric charge. This charge was concealed on Luftwaffe aircraft under the pilot's seat or near a vital control, like that of the rudder or ailerons. Two adjustments permitted explosion of the 500 grams (1 pound) of charge at 5,000 or 10,000 feet.

# "Limpet" Mark I Magnetic Mine

***Delayed Action Underwater Fuse or "A.C." Delay (Disassembled)***

a b c d

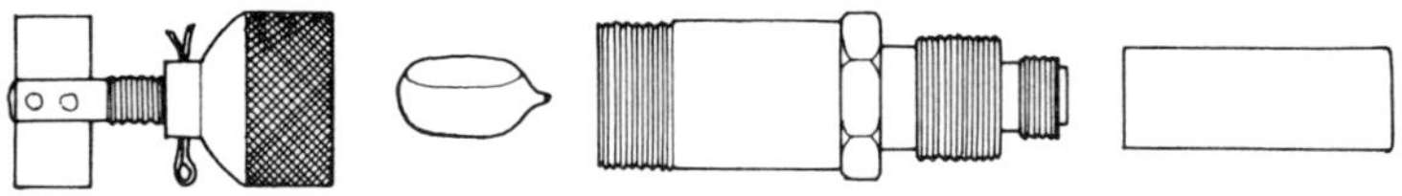

*Left to right:*

a. *Crushing screw protected by a safety pin*
b. *Bulb of colored acid*
c. *Fuse body containing the firing pin and primer*
d. *Special detonator to be screwed to the fuse body*

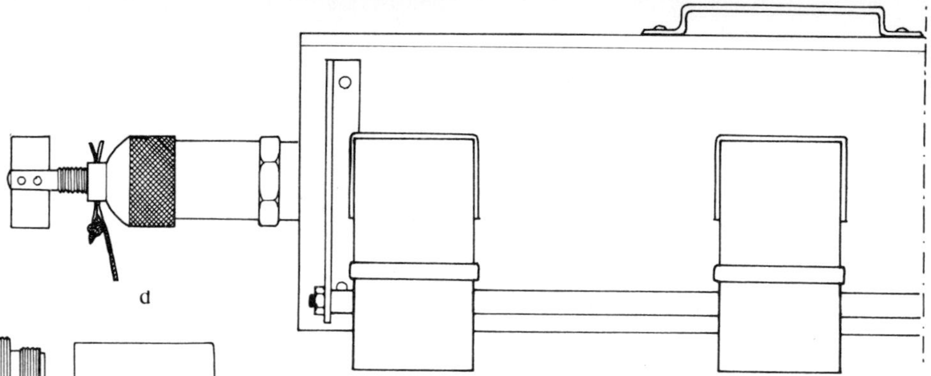

***Detail of the Limpet with the "A.C." Delay Assembled and Attached***

*To set the mine, all one needed to do was to remove the safety pin and to screw in the crushing screw, which broke the glass bulb.*

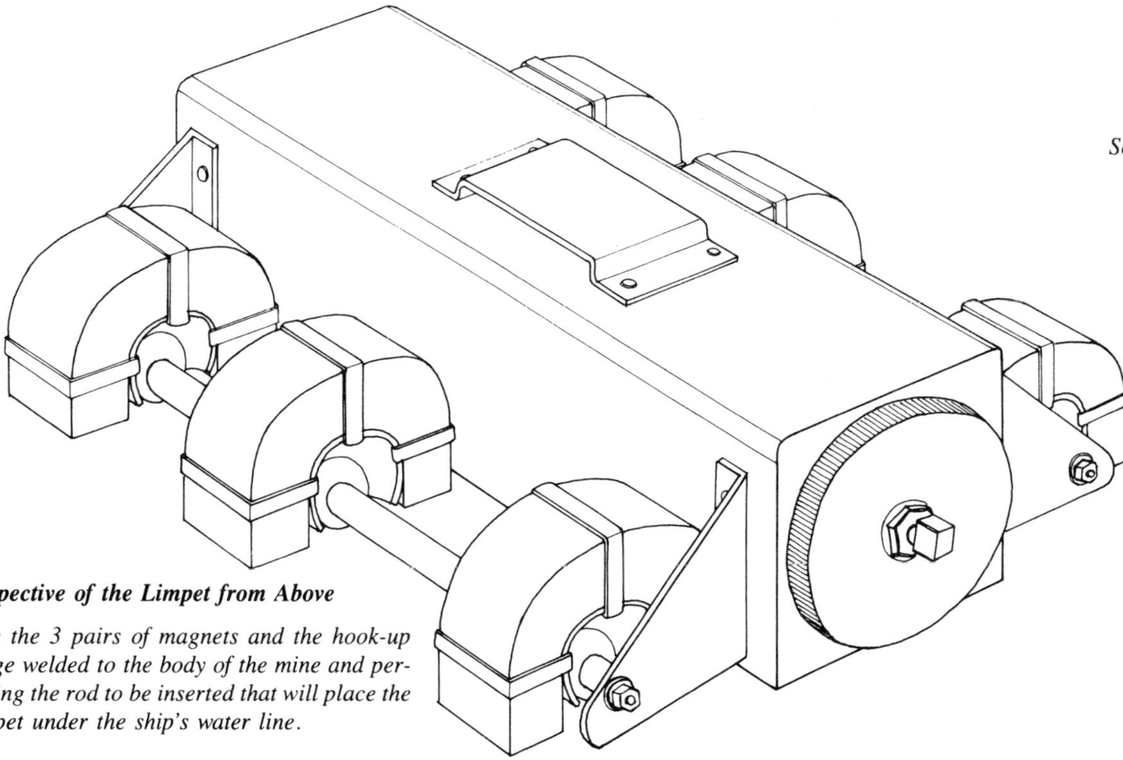

*Scale: approximately 1/3*

***Perspective of the Limpet from Above***

*Note the 3 pairs of magnets and the hook-up flange welded to the body of the mine and permitting the rod to be inserted that will place the Limpet under the ship's water line.*

NOTE: *At the end opposite the A.C. delay, there is a square screw head which can be removed to attach a second fuse called the "instant sympathy" fuse. Therefore, if several Limpets are attached to the same ship, the detonation of the first mine will induce a chain explosion of the others.*

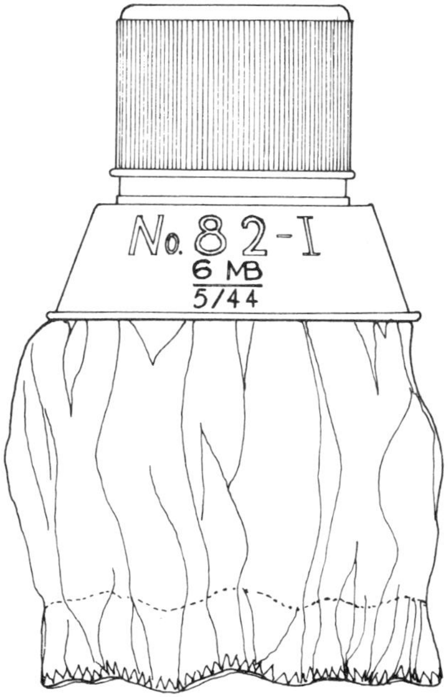


*An empty Gammon grenade with its black cloth skirt that holds the ball of plastic explosive.*

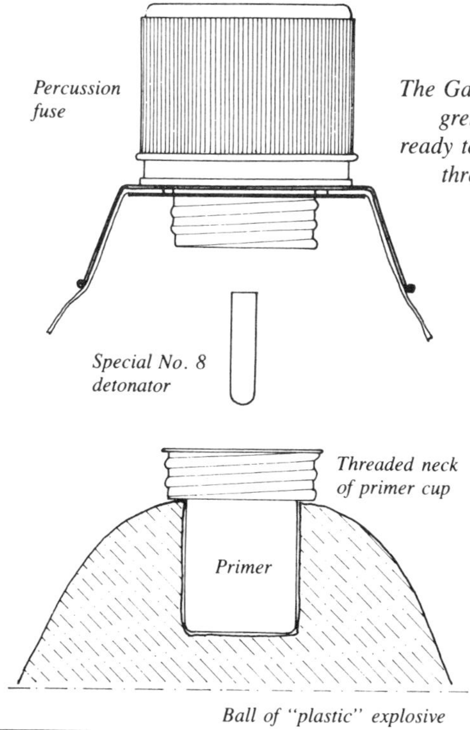


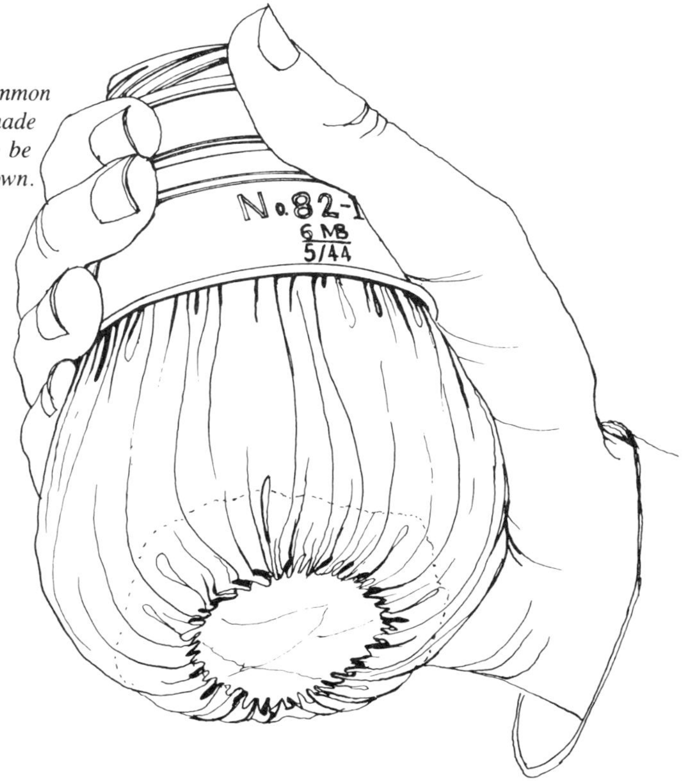


*The Gammon grenade ready to be thrown.*

*A loaded Gammon grenade with black bakelite cap unscrewed, weighted braid of a lead weight uncoiled, showing the safety pin attachment.*

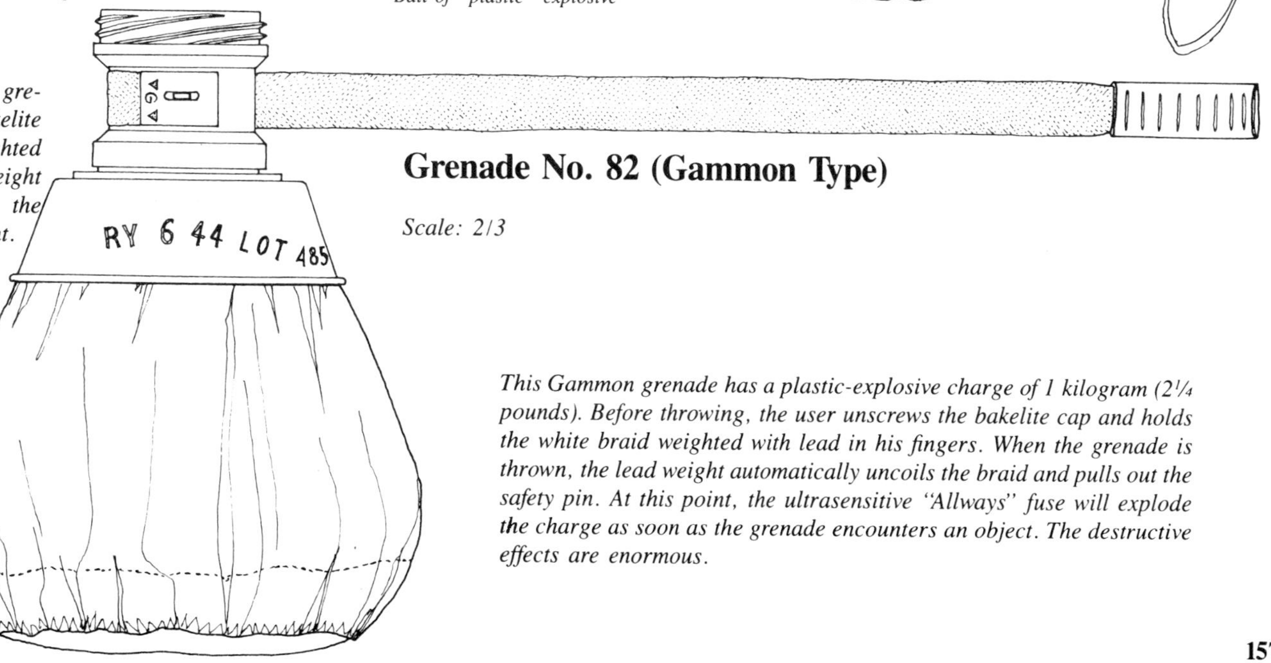


## Grenade No. 82 (Gammon Type)

*Scale: 2/3*

*This Gammon grenade has a plastic-explosive charge of 1 kilogram (2¼ pounds). Before throwing, the user unscrews the bakelite cap and holds the white braid weighted with lead in his fingers. When the grenade is thrown, the lead weight automatically uncoils the braid and pulls out the safety pin. At this point, the ultrasensitive "Allways" fuse will explode the charge as soon as the grenade encounters an object. The destructive effects are enormous.*

# Tyre Burster, Grenade No. 36 Mark I (Mills Grenade), and Fog Signal

*Scale: all 2/3*

***Tyre Burster***

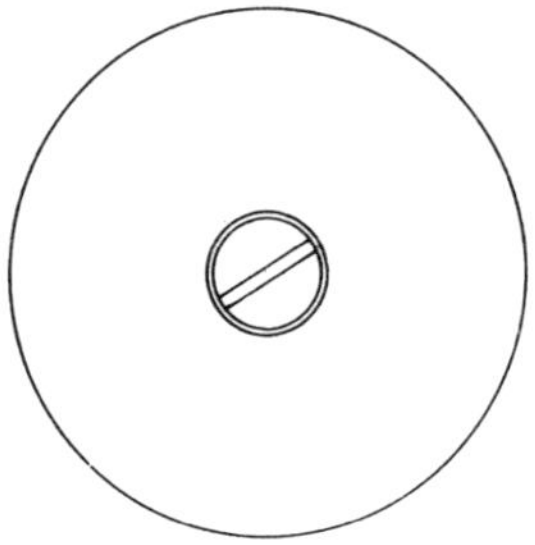

***Top View***

***Side View***

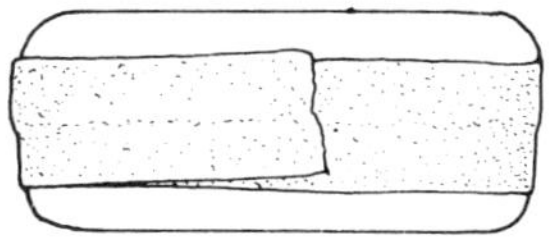

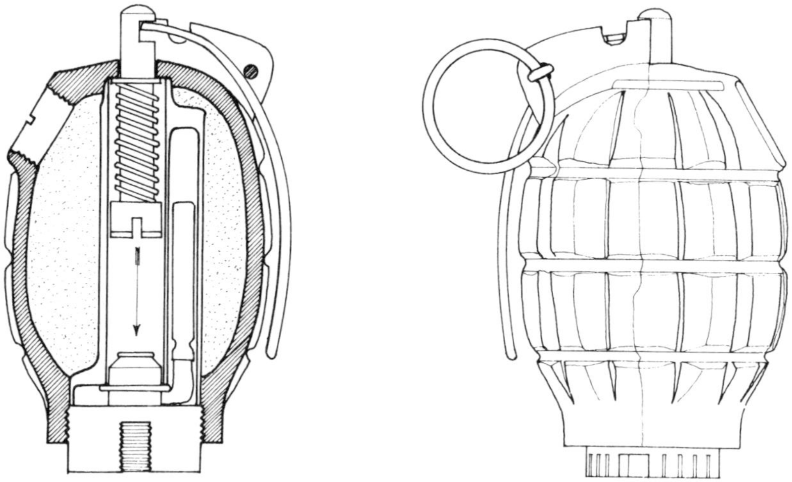

***Mills Grenade No. 36 Mark I***

*A cross section of the delayed-action igniter mechanism viewed from the side. The Mills is a defensive fragmentation grenade equipped with a 4-second delayed-action mechanism.*

***Fog Signal***

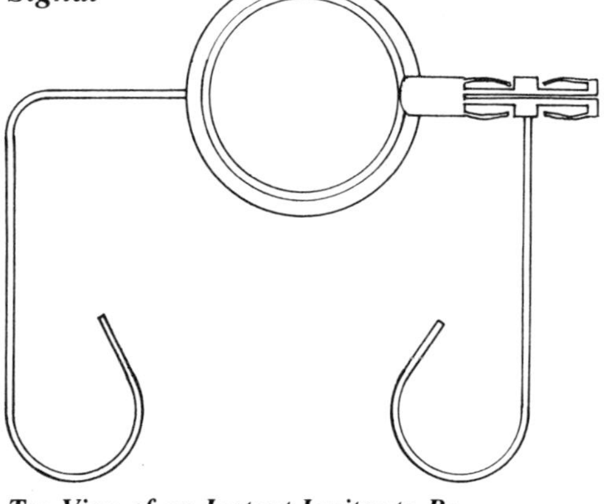

***Top View of an Instant Igniter to Be Placed on Railroad Tracks***

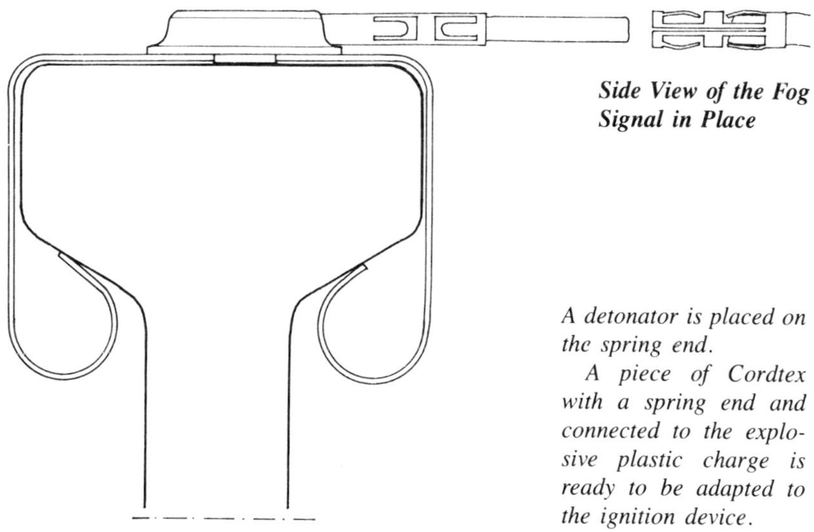

***Side View of the Fog Signal in Place***

*A detonator is placed on the spring end.*

*A piece of Cordtex with a spring end and connected to the explosive plastic charge is ready to be adapted to the ignition device.*

# Pocket Incendiary, Pencil Time Fuse, and Barometric Explosive



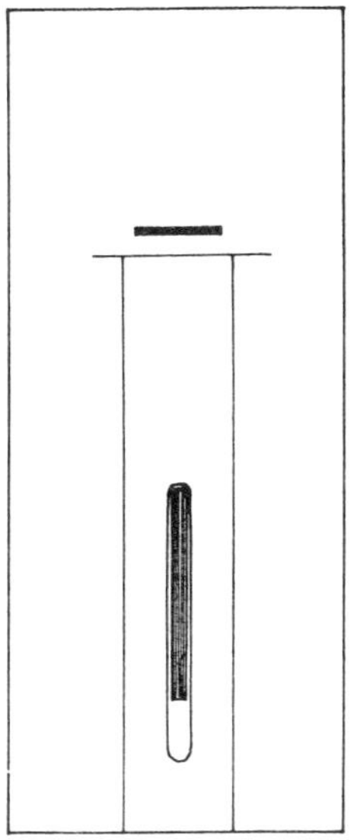

***Top View***

***Pocket Incendiary***

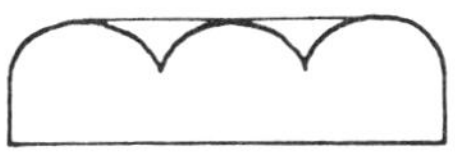

***Front View***

*Scale: 2/3*

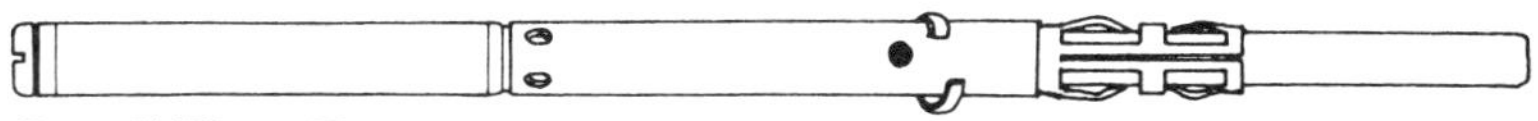

***Pencil Time Fuse***

*Scale: 2/3*

*Equipped with its detonator.*
*This is the acid bulb type.*
*Arming is accomplished by crushing the end of the tube.*

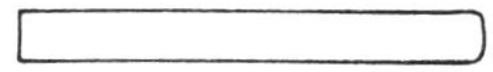

***Detonator***

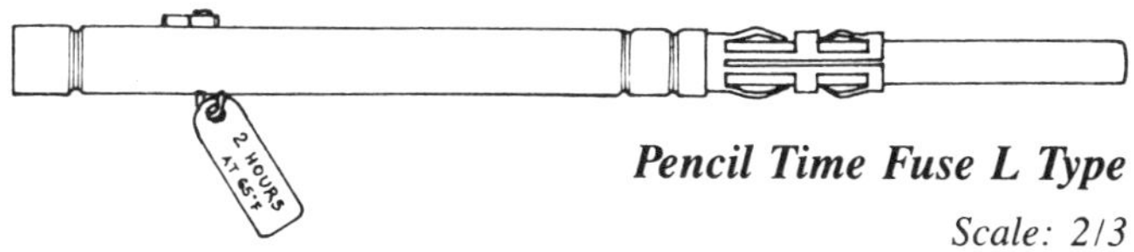


***Pencil Time Fuse L Type***

*Scale: 2/3*

*Equipped with its detonator. Type L without acid. Arming is accomplished by simply withdrawing the pin with the tag.*

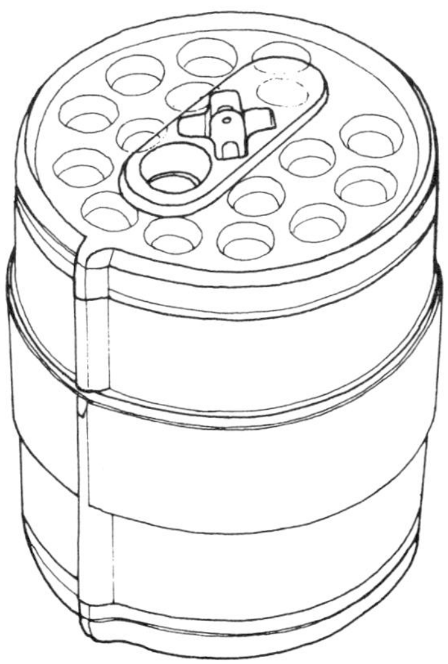

***Detonator Dispenser***

*Actual size*

*It could safely carry 16 detonators.*

***Barometric Explosive***

*Scale: 2/3*

*From left to right, the screw to set the explosion altitude, insulating plate with its safety pin, barometric bellows with electric contact, dry cell battery, electric primer, the detonator, primer, and part of the 40-centimeter (16-inch) rubber tube containing plastic explosive.*

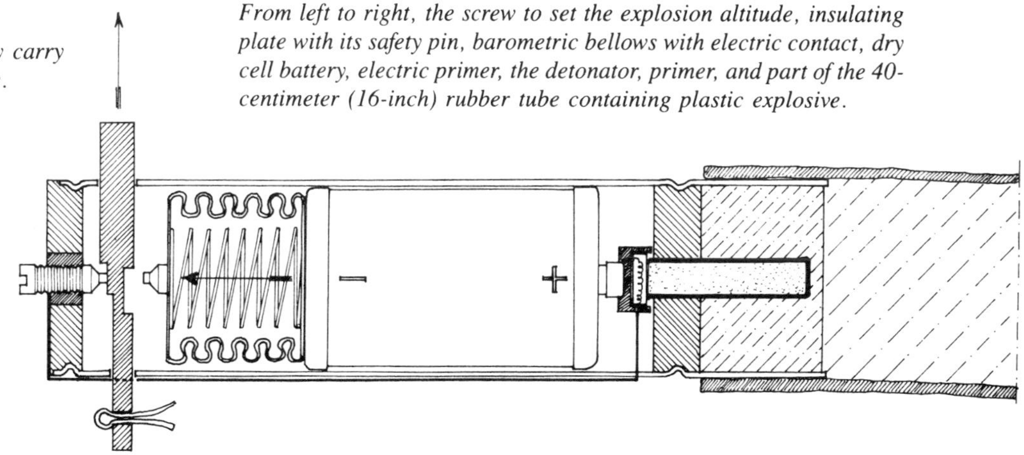

# CLOSE-DEFENSE WEAPONS

## Pen Pistol and Lapel Knife

### *Lapel Knife*

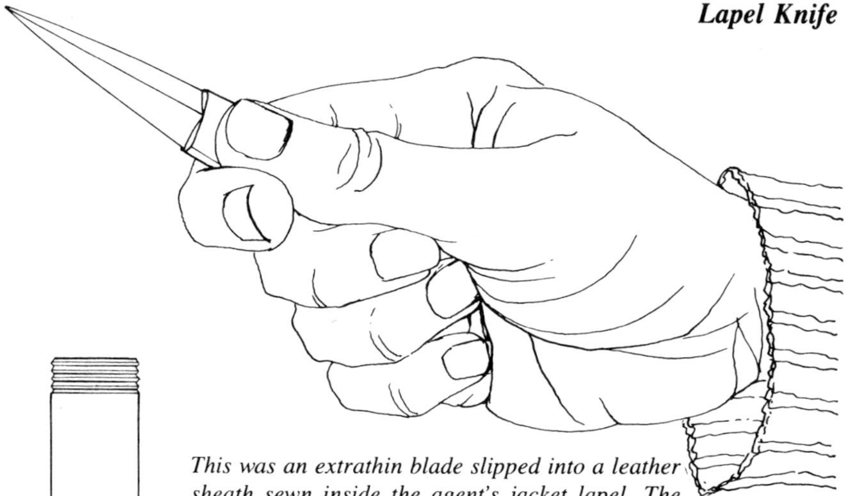

*This was an extrathin blade slipped into a leather sheath sewn inside the agent's jacket lapel. The mechanical action of the agent pulling out his wallet could turn into a quick stab to the face.*

***Blade Outside of Sheath***

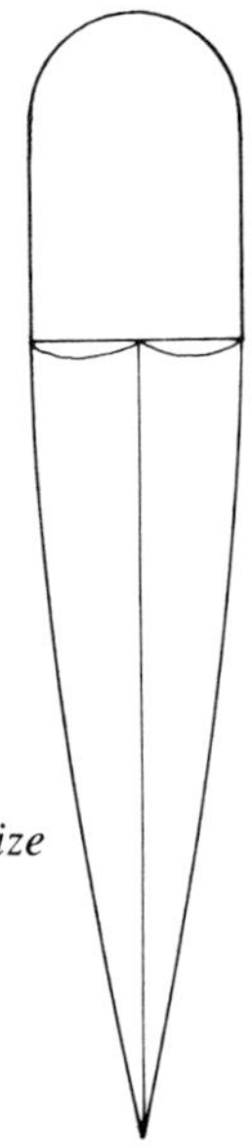

*Actual size*

***Front View***

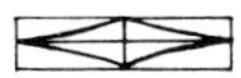

***Blade in Sheath***

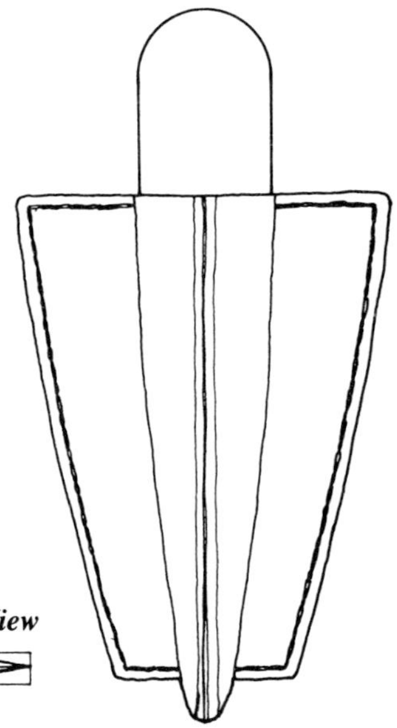

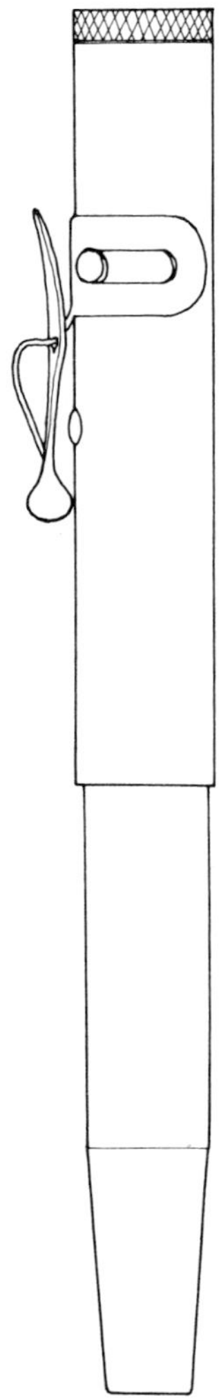

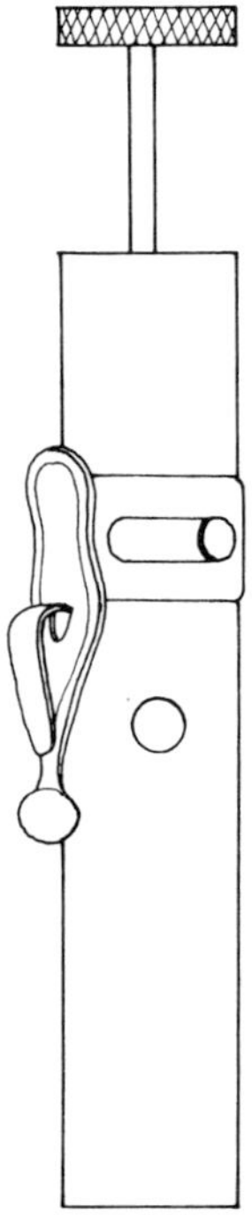

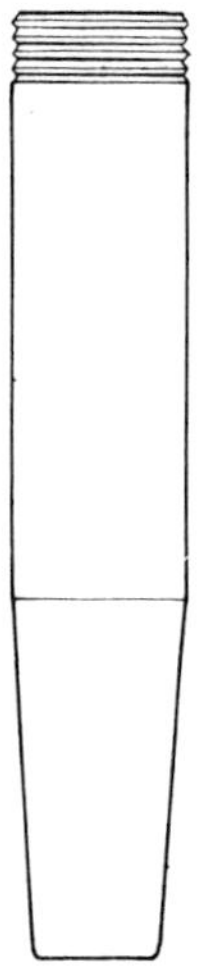

### *Pen Pistol*

*Actual size*

*This "pen" could be loaded with either a bullet cartridge or an asphyxiating gas cartridge.*

*The pen assembled and ready for operation. The clip serves as a safety device. Pushed to the right, it allows the ball tip to be depressed, triggering the shot.*

*The pen disassembled. The breech is open. The firing pin is armed by pulling the milled end back and then pushing it forward.*

*The cartridge is inserted into the chamber of the barrel, which is then screwed into the body.*

# Commando Dagger and Switchblade Knife

*Scale: 1/2*

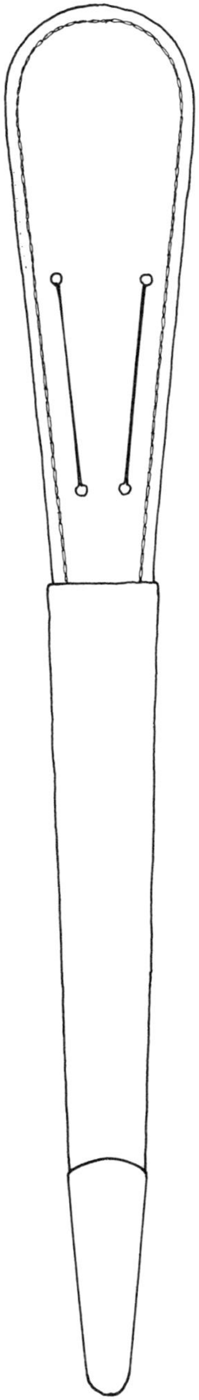

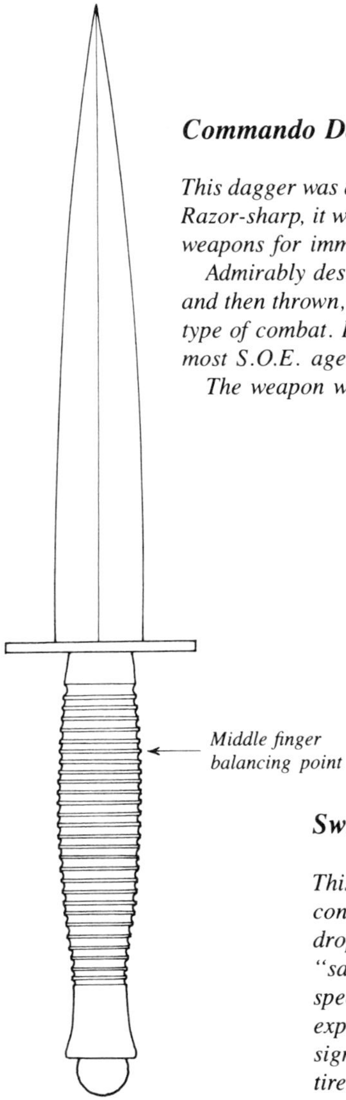


## *Commando Dagger and Its Sheath*

*This dagger was a regulation weapon used by British commandos. Razor-sharp, it was probably one of the quickest and most efficient weapons for immediate defense.*

*Admirably designed, it could be balanced on the middle finger and then thrown, and allowed for the usual tricks employed in this type of combat. It was very widely used as a personal weapon by most S.O.E. agents.*

*The weapon was completely blued dull black.*

## *Switchblade Knife*

*This knife was very sturdy and easy to conceal. A large quantity was dropped by parachute along with the "saboteur's knife." The latter was specially used to cut open slabs of explosives and had a short blade designed for puncturing and lacerating tires.*

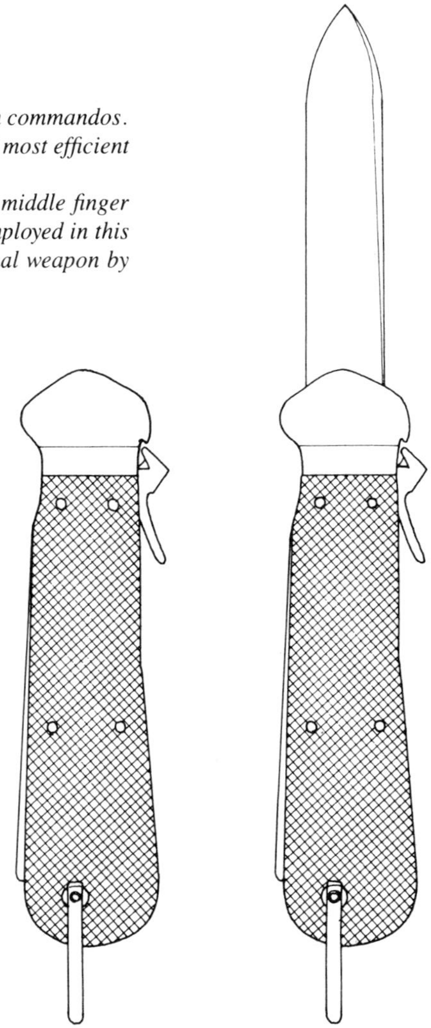

# Chapter VIII

# Epilogue

Such was the extraordinary clandestine armament the S.O.E. furnished to the Resistance.

The study of its evolution reveals a rapid and ingenious adaptation to the techniques of scientific subversive warfare.

By way of conclusion, it is important to briefly consider the negative consequences of these arms, on the one hand, and the positive results, on the other.

Underground fighter action correlatively entails reprisals. These are inevitable. They are part of the cost of total war and are hard on innocent populations only if one admits the principle of hostages.

The reprisals carried out by the occupying troops combined immoderation with horror. These reprisals quickly and surely brought the mass of the French population into the camp of active resistance.

The *Militärbefehlshaber* (military commander) declared: ". . . It is also my duty to hold the entire population responsible for the fact that up until now, we have been unsuccessful in our attempts to lay hands on these cowardly murderers and give them the punishment they deserve. . . . Among you are agents paid by the enemies of Germany."

Finally, to have a sufficient reserve of hostages on hand, the commandant of Greater Paris, after reminding everyone that all traffic was prohibited on public thoroughfares after the curfew, added: "German army patrols will make sure these rules are observed. Arrested offenders will be considered hostages."

The provocation-repression cycle was engaged, the Germans thus becoming, because of their insensitivity, the best recruiting agents of the Resistance.

The list of positive results is impressive and includes:

- Immobilization of factories and destruction of energy sources without losses to the civilian population.
- The economy of materials used for these missions of destruction. According to Professor M. R. D. Foot, the total weight of explosives that the S.O.E. used in France amounted to the weight of a single large-caliber airplane bomb, or 3,000 pounds.
- An economy of R.A.F. lives: a like result by aerial bombing would have resulted in the loss of many airplanes and their crews.

It is necessary to correct the opinion sometimes expressed that these sabotage missions were carried out haphazardly and had no military importance. In fact, each sabotage operation was carefully prepared and the subject of a formal and particular order transmitted by radio from the S.O.E. to the mission commander.

The objectives of sabotage missions varied according to strategical priorities and requirements drawn up by the Ministry of Economic Warfare: petroleum, submarine batteries, individual airplane parts, electronic devices, electric power stations, and for the months prior to the D-Day landing, all means of communication.*

But the most spectacular and probably the most beneficial result of the clandestine war was the arming and training of a secret army estimated at 150,000 men. There were, moreover:

- The delay of the SS 2nd Panzer Division (*Das Reich*) between Toulouse and the Normandy front by 15 days and of another armored division between Strasbourg and the landing beaches by 23 days.
- The diversion of 8 German divisions, out of a total of 60, for police operations.
- Finally, the decay of German morale resulting from terrorist attacks and sabotage of communication lines. On the eve of the D-Day landing, the S.O.E. gave, by means of "personal messages," the order to blow up rail routes and all lines of communication. More than 950 cuts were carried out that night, blocking almost all traffic in France but particularly depriving the German troops of their teletypewriter lines and compelling them to resort to radios to pass their orders. Within 48 hours, Allied monitoring stations were able to break the German messages and so learn of the enemy's intentions and future movements.

One of the most effective tactics of the Allied High Command was to order S.O.E. missions to blow up

*We have held in our hands a telegram from London to an F Section mission commander, listing sabotage operations to be carried out: 2 airplane construction factories, a rubber reconstituting factory, a tungsten mine, and a French railroad depot.

bridges on secondary roads so that retreating German motorcades would be forced to use primary roads. Teams on the ground radioed assault planes, which destroyed the vehicles. The more experienced German troops then began again on the secondary roads, often on foot, where they fell into ambushes set up by the Maquis (French Resistance fighters).*

The Allied Command estimated the combined action of the S.O.E. and the clandestine army it supplied as having reduced the overall length of the war by approximately 6 months.

Accordingly, subversion appears to have been one of the most effective and economical weapons of the arsenal of total war. Experience proved that it could be directed only by a secret service attached to the highest echelon of command and that it could only succeed in a favorable area which was methodically and patiently prepared, over a long period of time, down to the last detail.

If the S.O.E. was successful in France, it was thanks to the active or passive complicity of the vast majority of the population and because of the spirit of resistance it activated.

Without the S.O.E. as its tutor, the French Resistance would have been doomed to suffocation and defeat.

Five hundred British officers in F Section (460 men and 40 women, according to Colonel Buckmaster) volunteered to serve in France.

Each was warned that he or she had 1 chance in 2 of returning alive.

More than 130 from this division alone were killed in combat, shot, hanged, or poisoned in concentration camps.

To all those who chose the strait gate to help our country in distress, and who, like brothers, shared for four years their lives and their daily perils with us, goes our profound gratitude.

*For example, on August 9, 1944, in the zone of the S.O.E. FIREMAN circuit and the "Jedburgh" mission ALEXANDER, a British lieutenant with 5 French fighters surprised 35 German bicyclists. Of these, 21 were killed and 12 wounded. The Allied team withdrew without losses after an engagement of less than a minute.

# Appendices

# Radio Authentication Systems

The intercession of eminent specialists at the end of 1942 revolutionized the S.O.E. communication system.

Let us summarize the important decisions made during the second half of the war:

- The abandonment of the impractical and not very safe double transposition system and the introduction of the mathematically unbreakable one-time pad.
- The abandonment of complex systems of authentication (security checks); introduction of systematic checks of radio fingerprints of the agents in the field by the TINA system.
- The abandonment of rigid and dangerous calling schedules and the introduction of the mathematically determined "mosaics" system.
- The abandonment of the principle of dialogue between the home station and the network and the introduction of naval broadcast procedure whereby transmission and reception are separated in time and space.
- The depersonalization of communication from the home station through the exclusive use of automatic telegraphy.
- The establishment of the "electronic umbrella."

We looked at some of these decisions in the chapters on transmission and codes. To complete the picture, we will make a quick analysis of security checks, the TINA system, mosaics, and the electronic umbrella.

## Security Check Systems

Although the correct origin of telegrams was verified by call signs (which were really nothing more than passwords), it was also imperative that an agreed-upon signal assure the home station that the agent was not operating with a gun at his back.

With ciphers that could be solved, it was impossible to insert an agreed-upon signature within the text itself of the enciphered telegram. The Germans would have ended up figuring out the pseudonym, and since it had to be repeated in each telegram, it would have rendered their cryptanalysis easy and rapid.

The security check, or security-group principle, consisted of inserting the signature in the first group preceding the ciphertext and of repeating it, in another form, in the last group.

It goes without saying that this pseudosignature had to change frequently, which created an extra burden on the radio operator's memory.

Suppose that BOY is the signature which the agent uses to show he is operating freely. These 3 letters are camouflaged in the first group by replacing each letter with the second letter following in alphabetical order: BOY becomes DQA. Any 2 letters could be added to get a 5-letter group: for example, SDQAT.

In the last group, the pseudonym is camouflaged by replacing each letter with the letter following fifth in alphabetical order: BOY becomes GTD. Any 2 letters are added to complete the group: KGTDA.

The security check system was, furthermore, backed up by a series of test questions which the home station could ask in case of persistent doubt. Thus, if, to the question "Do you want a shipment of chocolate?," the radio operator gave the logical affirmative response, he was no longer free. The agreed-upon reply of "My pajamas are blue" meant that the network was operating normally.

In trying to guarantee too much, the system of the initial years became completely snarled. Little by little, the agents and the home stations no longer paid attention to it. Then, that which had to happen, happened. This was the catastrophe of Section N (Dutch), in which most of the S.O.E. radios were intercepted and turned by the Abwehr as they were dropped by parachute.*

A well-qualified, upper-echelon official has willingly confided that his opinion was, and always will be, that the security-check system was a double-edged weapon that could have been turned against the S.O.E., to the immense harm of the latter.

At the end of 1942, a drastic decision was made concerning the sacrosanct security-check and double trans-

*This was often called the "North Pole" operation. This name is misleading and requires explanation. The English called the Dutch S.O.E. section the "N Section." The Germans also designated this group with that initial. Now, in telephonic procedure in the Wehrmacht, the letter "N" was customarily phonetically given as "Nordpol" (North Pole). So a literal translation of this word has no real meaning.

position systems. The decision was to do away with them immediately.

As the F Section missions were being progressively supplied with the one-time pad, it was decided to incorporate the agreed-upon pseudonym signature into the text itself of the enciphered telegram. In this way, the radio operator only had 2 signatures to remember: one which indicated that he was free, and another which meant he had been taken prisoner. Perhaps he would only need to remember the first if he took the escape route offered by his cyanide capsule. This ultimate sacrifice guaranteed the network's survival. If the Germans then committed the imprudence of operating the small black suitcase, they opened the door to the poisoned Machiavellian dialogue of the Intelligence Service, to the Allies' great benefit.

For the TINA recording system confirmed the operator's identity beyond a shadow of a doubt.

## TINA Radio Fingerprint Analysis

We give an example below. Before leaving, each radio operator left a sample of his Morse "writing" by recording all the letters of the alphabet and all the numbers. Marking them on a paper tape that moved at 16 feet per minute permitted an extremely precise analysis of signals, for a Morse dot was a quarter of an inch long on the tape.

The irregularities of dots, dashes and spaces characteristic of the individual's operation style could easily be made out on the record.

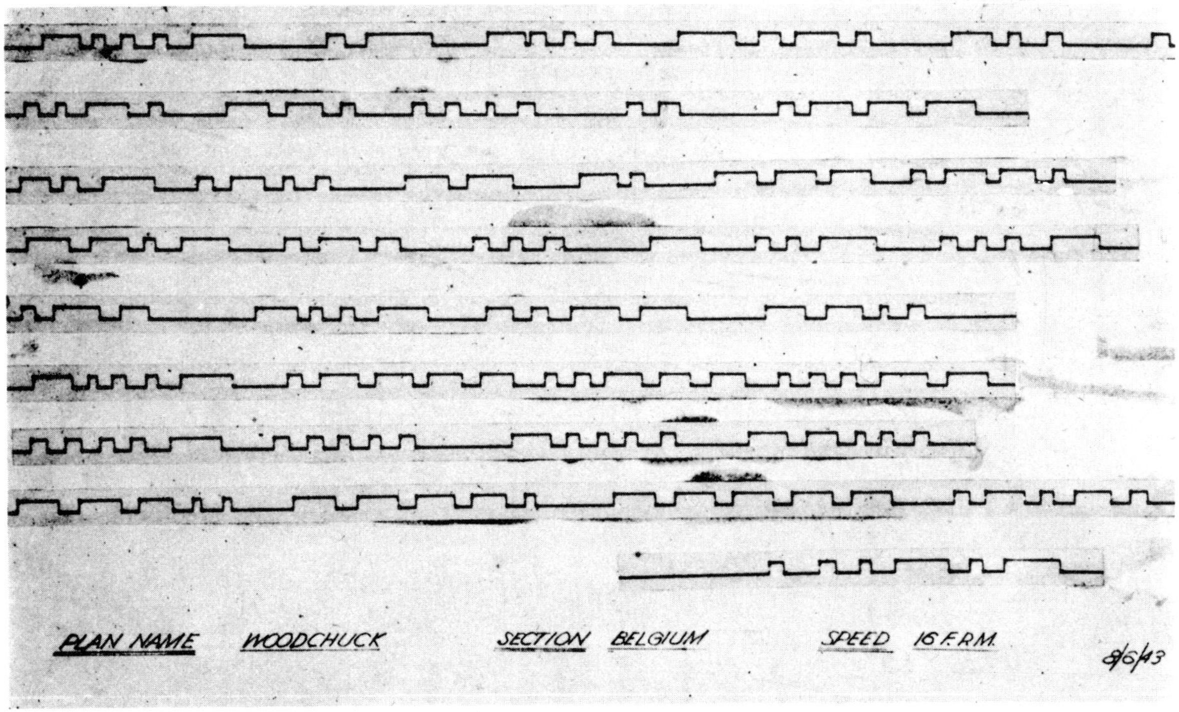

Nevertheless, to have a more schematic visual representation and for instant identification, the length of dots, dashes and spaces for each letter was measured with special transparent rulers and transferred in graphic form to square-ruled paper.

Shown are 2 graphs drawn up after recording agent Mackintosh's signals:

It is immediately obvious that these 2 transmissions are not from the same hand.

The principal characteristics of the Mackintosh of July 9, 1943, are long spaces between the dots and dashes, and very short dots at the end of letters ending with a dot, such as *f* and *h*.

The Mackintosh of January 11 and 15, 1944, is characterized by excessively long dots for letters ending with a dot (*f*, *h*, *i*, and *l*), and a long dash in letters ending with a dash.

We should reassure our readers by informing them that the second Mackintosh was not an Abwehr operator, but rather a second operator which the first Mackintosh enlisted to help.

The TINA system operated smoothly from the summer

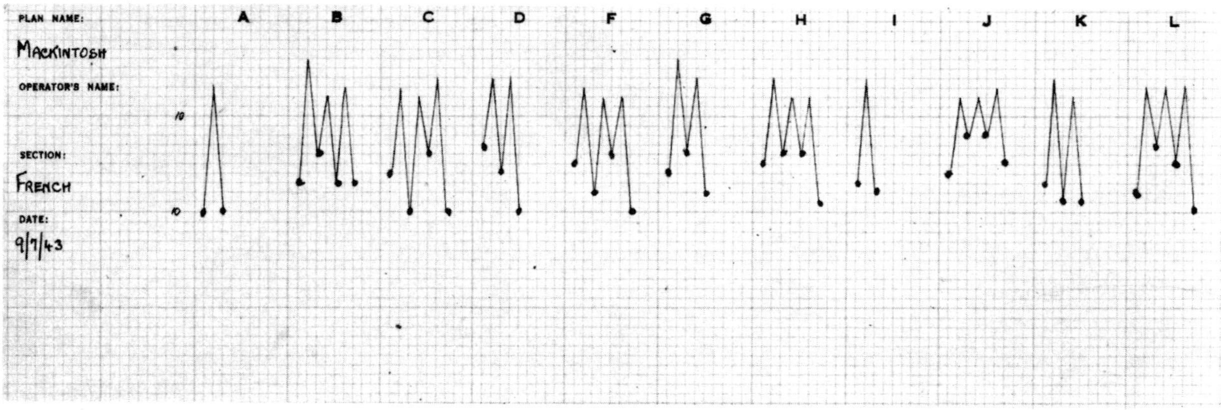


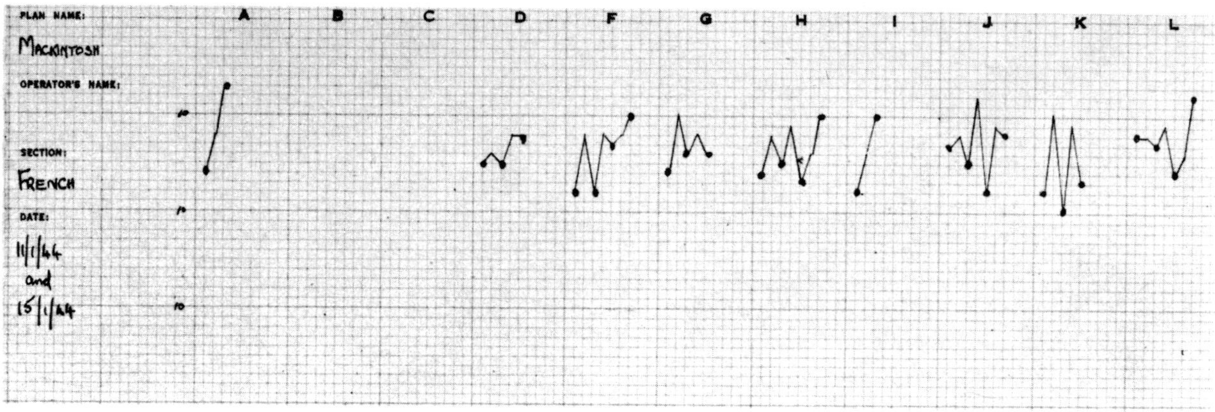

of 1943 until the end of hostilities. More than 600 individual cases were examined. Verification recordings were carried out every month.

No matter how much nervous tension or anxiety the clandestine operator worked under, his Morse code signals never lost their general characteristics and the recording reproduced them with precision.

Each time there was a change in operator, it was immediately detected. When, after the war, the radio operators were informed that their identities had always been determinable, some expressed the opinion that if someone had explained before they left just how effective the system really was, their nervousness would have been considerably lessened while they were operating.

### The "Mosaics"

During the first half of the war, each operator had a microfilmed calling schedule which strictly established the code signals and frequencies to use, as well as the days and times for his sessions. This method probably caused some losses. In any case, it made the task of the enemy's interception services easier.

Therefore, a decision was made to entrust a team of experts with the job of coming up with a "grand mosaic" giving the same communication times, the same frequencies, and the same call signs to stations belonging to networks which were geographically far apart from one another.

So, the German interception services, listening every day at the same time and on the same frequency to a station communicating with the same call sign, were, in reality, hearing, by the end of a week, seven different networks.

Furthermore, each individual "mosaic" was set up in such a way that no single station had any logical repetition of times, frequencies, days, or call signs for the period of its service.

Part of the "grand mosaic" of Home Station 53-B, Channel 45, is shown here. This was set up for one year and controlled 13 networks.

For example, on January 10, between 14:45 and 15:45

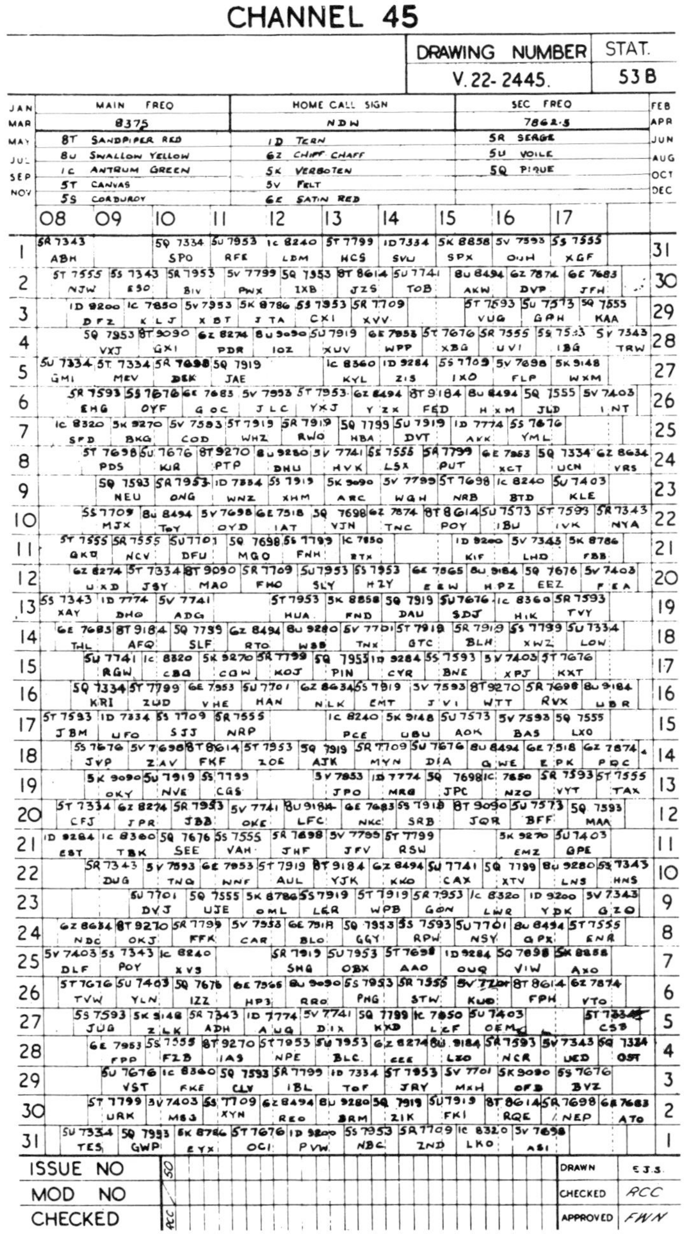

Greenwich Mean Time, the frequency of 8,614 kilocycles per second is reserved for the SANDPIPER RED network with the call sign POY.

### The Electronic Umbrella

A very experienced ear could recognize the particular tonality or sound of signals transmitted by the secret sets we have described. This is due to the fact that these transmitters were meant for giving out maximum power with minimum size and were not provided with the necessary number of circuits for obtaining perfect tone quality.

Therefore, although the B2 and B.P. 3 sets were stabilized with quartz crystals, their characteristic tones could not fool the Germans about their identities.

To minimize this danger, and before setting up a new network that would use Polish B.P. 3 sets, for example, the English secret services installed a phantom network on British soil, using a number of such sets. This network operated on the frequencies and with the same call signs as the future network, but the communications transmitted were all fake.

The German interception services listened in on this phantom network and localized it. Once they realized that all the transmitters were located in English territory, they abandoned their scrutiny of the network.

At this point, the S.O.E. took sets out of the phantom network as needed, dropped them by parachute into the occupied territory, and added them to the real clandestine network.

In this way, the German radiolocation services were slowly but surely led into saturation and chaos.

## Results of German Direction-Finding

An analysis of the official British history, *S.O.E. in France*, enables us to set down a table summarizing the number of F Section stations captured specifically as a result of German radio direction-finding, the reasons for some of the captures, and the dates on which they took place.

| *Date of Capture* | *Station Location* | *Reasons for Capture* |
|---|---|---|
| November 1941 | Le Mans | Transmission too long and from the same site. |
| October 15, 1942 | Lyons | Captured during too long a transmission. |
| October 24, 1942 | Lyons | Captured during too long a transmission. |
| December 1942 | Lyons | Captured after 7 months of communications. |
| July 31, 1943 | Pontoise | Not specified. |
| November 29, 1943 | near Nantes | Indiscretion of liaison officer transporting the set. |
| January 15, 1944 | Saint-Quentin | Transmissions from the same site during several months and an operation plan showing periodic repetition of days, hours, and frequencies. |
| July 1944 | Lyons | Captured after 5 months of communications. |
| July 1944 | Normandy | Transmission of immediate intelligence while agent was too near German units. |
| July 1944 | Paris region | Not specified. |

If one compares the number of captured stations to the number of stations in service, the ratio of losses can be set down as follows:

1941: 1 capture, or a ratio of 1 to 7 stations in operation.
1942: 3 captures, or a 1:7 ratio.
1943: 2 captures, or a 1:11 ratio.
1944: 4 captures, or a 1:13 ratio.

In all, 10 captures exclusively from enemy direction-finding out of 100 network stations.

Five of these 10 captures were due to carelessness and the nonobservance of basic security rules. Finally, 2 of the captures would have undoubtedly been avoided if the transmissions had not taken place in the same town where one of the regional bases of German direction-finding was located.

Therefore, while recognizing the astounding technical success of the automatic position-finding system, one should not *overestimate* its successes.

Many operators, like other network members, were captured because of thoughtless chatter, indiscretions, police investigations, or just routine road and railway checks.

Finally, it is said that happy people have no history, and the 10 captured operators should not cause us to forget the extraordinary usefulness and effectiveness of the remaining 90, to speak only of F Section.

To cite only one case, not belonging to the S.O.E., the French KLÉBER intelligence unit never had even the least incident throughout its transmission history. It had 8 operators working with Algiers from the immediate environs of Pau, near the border with Spain. In 1944, the transmissions of this network did not exceed 3 minutes for each frequency.

# Layout of Transmitter Aerials at Secret Station 53-B

Isolated in the English countryside in the middle of a 125-acre field, watered by a small stream and surrounded by tall trees, the station looked like a spider in the middle of its web.

This web consisted of 34 folded dipole aerials and 7 high-gain rhombic aerials.

In the middle of the diagram is the concrete base on which 2 prefabricated buildings were erected. The larger of the structures sheltered about 50 transmitters, which could all be operated at the same time. Each transmitter had an antenna power of 250 watts. A large 15-kilowatt transmitter was kept on hand in case of bad propagation conditions. A small building inside an extension of the transmission room housed the power station and the auxiliary generators.

An access road was laid out.

The number of control technicians was extremely low since all the transmitters were remote-controlled by operators located in the camouflaged receiving station, which was isolated and far removed from the transmission station for elementary security reasons.

Note, on the layout, the arrows, accompanied by a magnetic deflection, indicating the direction in which radiation is optimal. These show that 30 aerials emit their maximum radiation toward the east, 4 toward the southeast, 4 toward the northeast, and 3 toward the south.

For long-distance communications, rhombic aerials were used, which, depending on the dipole, provided a gain of 8 to 10 decibels. They were directed toward Yugoslavia, Greece, and the Middle and Far East.

On the right side of the field, the 4 dipole antennas marked 60° are directed at agents in Norway.

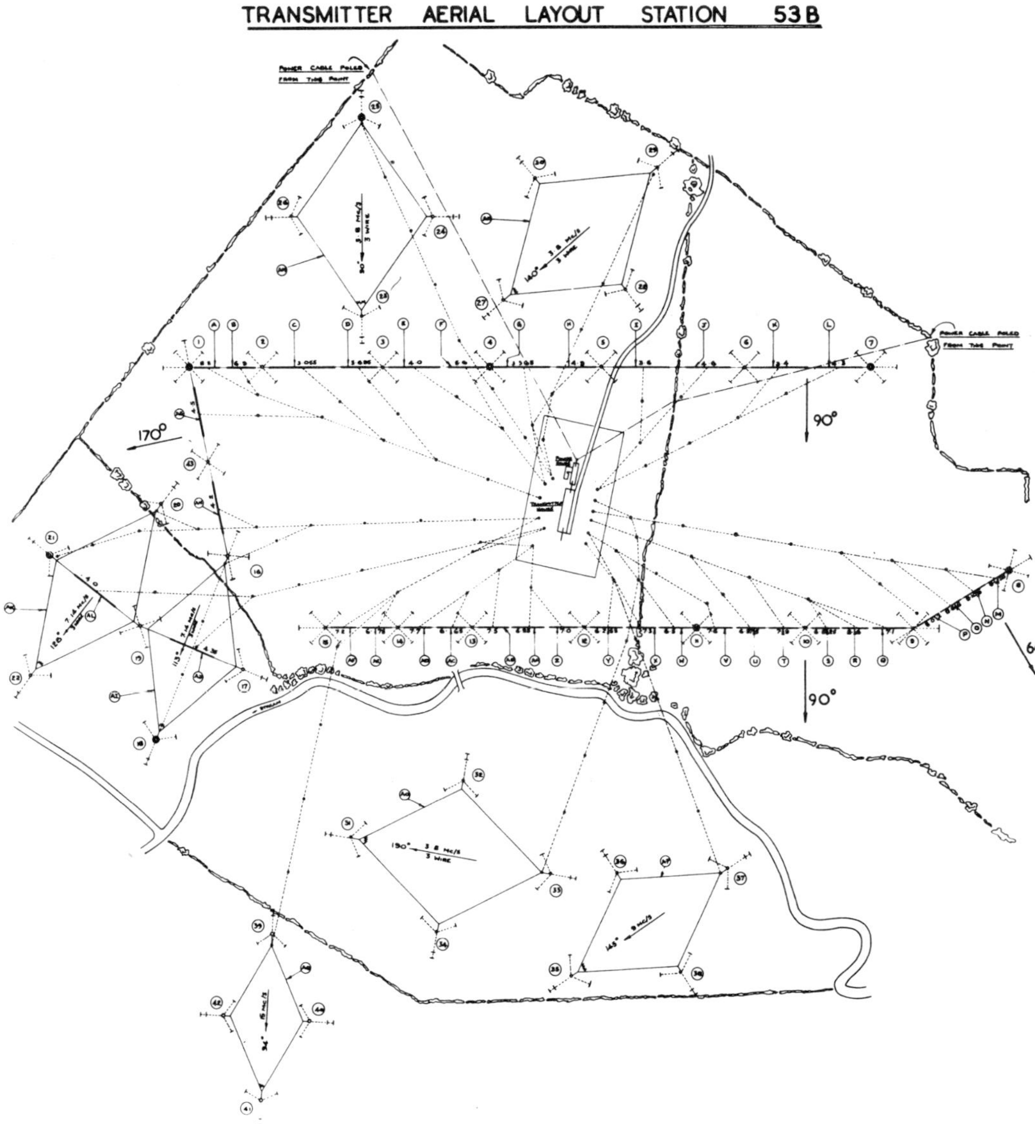

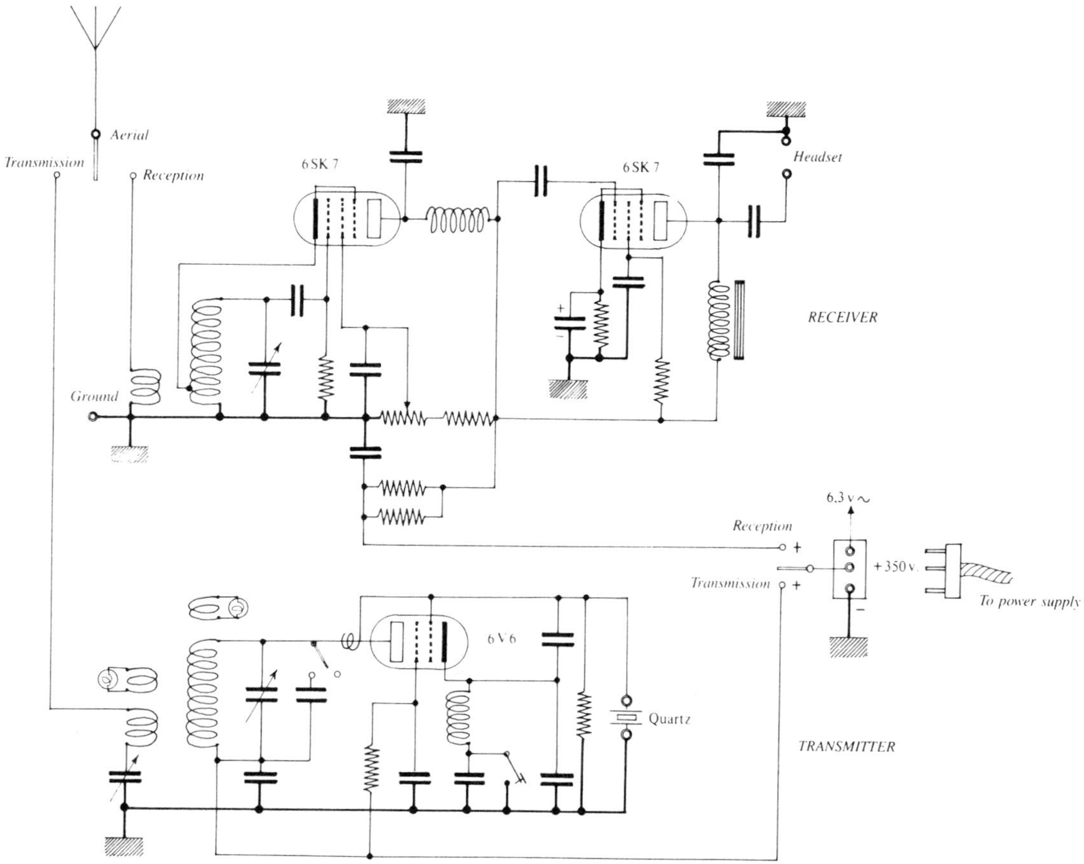


# Circuit Diagram of the Paraset

The receiver is made up of 1 cathode reactive detector stage (6SK7) and 1 audio-frequency amplifier stage (GSK7).

The transmitter has a single quartz crystal oscillator stage (6V6). The Paraset's excellence lies in its extreme simplicity, which guarantees, with an absolute minimum of components, easy repair and maximum compactness. On the other hand, its receiver is sensitive but unstable, and selectivity is very poor. The regeneration control is noisy and induces an oscillation in the aerial that disturbs all broadcast receivers within a hundred-yard radius. Accordingly, receiver operation alone could attract the attention of the German mobile listening services.

For this reason, whenever possible, Parasets were removed from the city networks and used in the field where receiver radiation would have less dangerous consequences.

The Paraset and Mark XV receivers were analogous. The latter, however, had an auxiliary tube which prevented receiver oscillation from reaching the aerial.

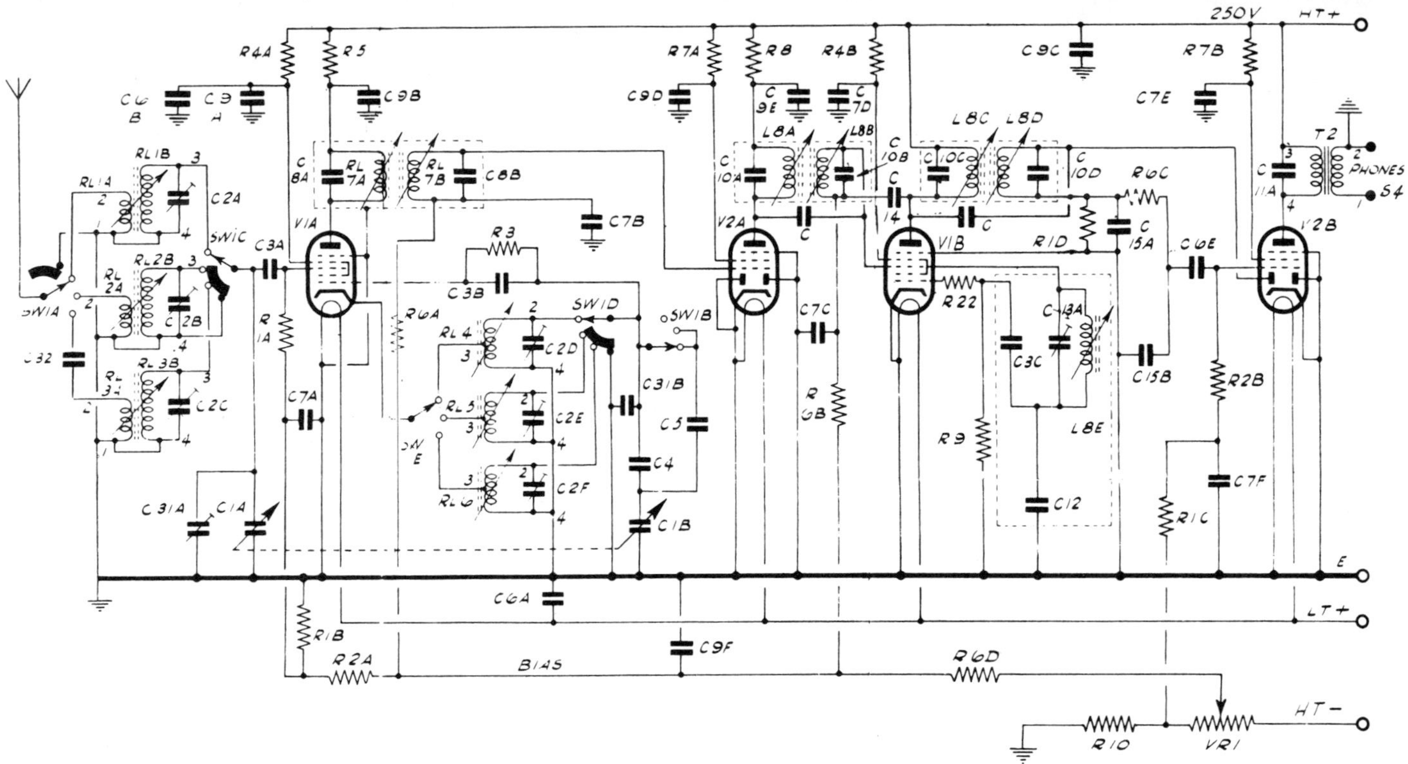


| RESISTORS | | R6A | 150K 1/10 WATT |
|---|---|---|---|
| R1A | 1 MEG 1/10 WATT | R6B | 150K " |
| R1B | 1 MEG " | R6C | 150K " |
| R1C | 1 MEG " | R7A-B | 100K 1/4W. INSULATED |
| R1D | 1 MEG " | R8 | 1K 1/2W WIREWOUND |
| R2A | 4 MEG " | R9 | 20K 1/10 WATT |
| R2B | 4 MEG " | R10 | 40K " |
| R3 | 20K " | R22 | 50 OHMS " |
| R4B | 20K 5% WIREWOUND | VR | 100K POTENTIOMETER C.T. LOG LAW |
| R4A | 20K 5% WIREWOUND | | |
| R5 | 5K. 1/2W. WIREWOUND | R6D | 470K 1/10 WATT |

| COILS | |
|---|---|
| RL1A-B | AERIAL COIL 8·7-15·2 MCS |
| RL2A-B | " " 5·2-9·04 " |
| RL3A-B | " " 3·1-5·4 " |
| RL4 | OSC COIL 8·7-15·2 " |
| RL5 | " " 5·2-9·04 " |
| RL6 | " " 3· -5·4 " |
| RL7A-B | IF COIL 470KC 7/45 LITZ |
| L8A-E | IF COIL " " |

| CONDENSERS | |
|---|---|
| C | 2p +0-1p CERAMIC |
| C1A-B | 2 GANG 90p |
| C2A-F | TRIMMER 2-9p K 02509 |
| C3A-B | 100p 10% CERAMIC |
| C3C | 100p 10% CERAMIC |
| C4 | 700p -2% +1% MICA |
| C5 | 600p -0% +3% MICA |
| C6A-B | ·001 250V MICA |
| C6C | ·001 250V MICA |
| C7A-B | ·01 250V MICA |
| C7C-D | ·01 250V MICA 10% |
| C7E-F | ·01 25% MICA |
| C8A-B | 110p ± 2p MICA |

| | |
|---|---|
| C9A-F | 0·1 PAPER 10C/4021 |
| C10A-D | 220p ±4p MICA |
| C11A | ·002 350V MICA |
| C12 | ·001 ±20p MICA |
| C13A | 25p VARIABLE |
| C14 | 6p ±20% CERAMIC |
| C31A-B | 6-8p CERAMIC |
| C32 | 50p ±10% |
| C15A-B | 100p ±20% MICA |
| VALVES | |
| V1A-B | 7Q7 |
| V2A-B | 7R7 |

## Circuit Diagram of the Receiver of the 3 Mark II ("B2") Suitcase Transceiver

This was, without a doubt, the best clandestine receiver employed by the British services. It was equipped with 4 "lock-in" multiple tubes. It had 1 frequency converter stage (7Q7), 2 intermediate frequency stages on 470 kilocycles, 1 beat frequency oscillator stage (7R7 and 7Q7), and 1 detector and audio-frequency amplifier stage (7R7). Its sensitivity was 1 to 3 milliwatts for 10 milliwatts of output at 1,000 cycles. The selectivity was 2 kilocycles at 6 decibels and 18 kilocycles at 40 decibels. Power consumption was 9.5 amperes at transmission, key depressed, and 4.5 amperes at reception.

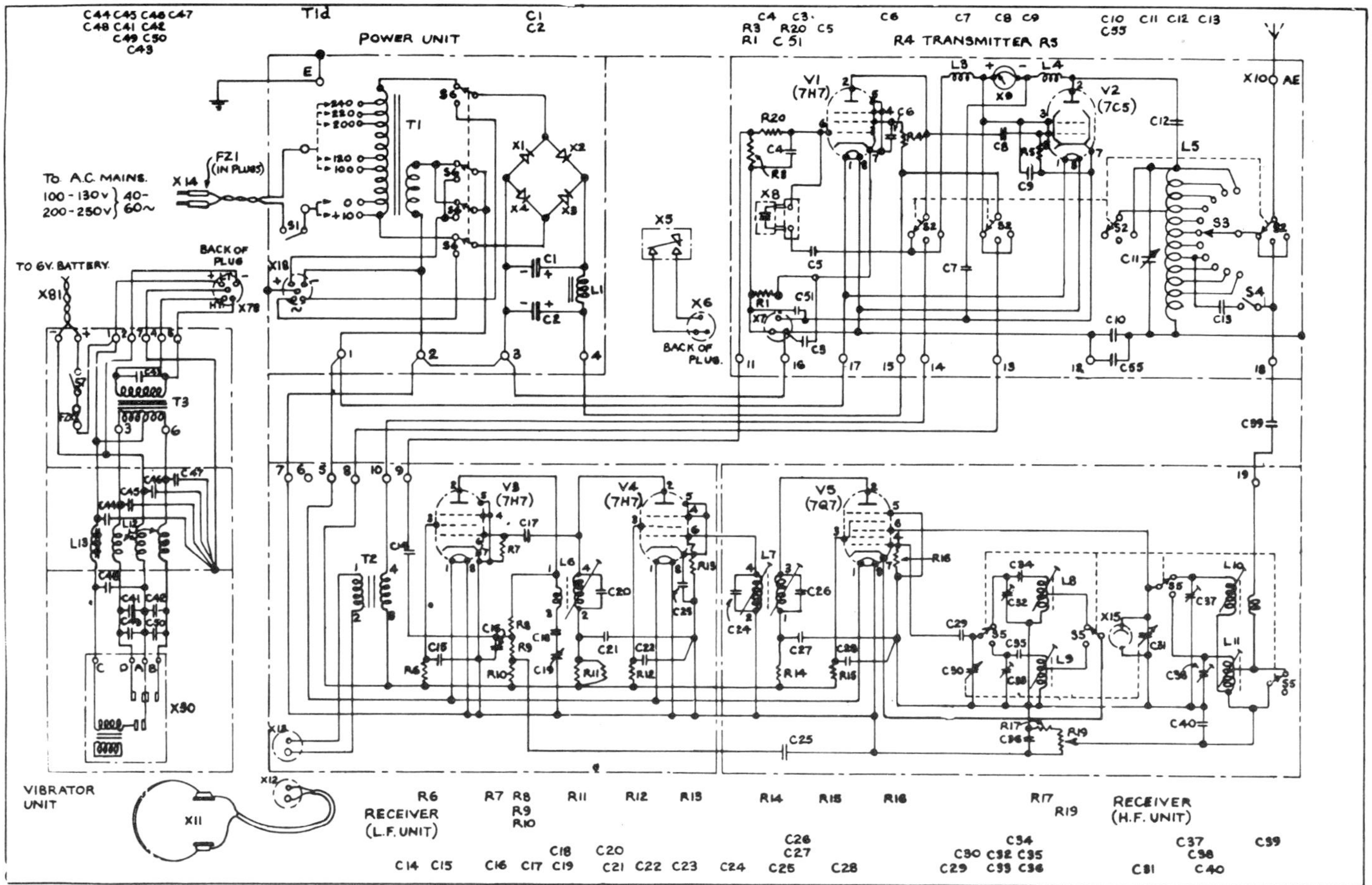


# FIG. 7

BASE SPARES AND DIAGRAMS
(THEORETICAL CIRCUIT DIAGRAM)

REV A: C51 WAS MARKED C1 P.C.L.003/194. 19·12·44

C.D. 2043

## Circuit Diagram of the A Mark III Suitcase Transceiver

The set was equipped with 5 American "lock-in" tubes which fulfilled 7 functions. When the set was on "receive," the quartz-controlled pilot tube of the transmitter served as an audio-frequency amplifier. The receiver (lower circuit) was comprised of 1 frequency converter stage (7Q7), 1 intermediate frequency stage at 1,200 kilocycles (7H7), 1 reaction detector stage which increased sensitivity and selectivity and permitted the reception of Morse Code signals (7H7), and 1 audio-frequency stage (7H7). The transmitter (upper circuit) had a crystal-controlled pilot stage (7H7) and an amplifier stage (7C5). The power supply unit with its oxymetal rectifiers eliminated the rectifier tube still found on the Polish A.P. 4's. Receiver selectivity was 2 kilocycles at 6 decibels and 20 kilocycles at 40 decibels. The sensitivity was better than 10 microvolts per 1 milliwatt with a load of 600 ohms. The signal-to-noise ratio was about 10 decibels. The receiver was stable enough to accept a 400-cycle audio-frequency band filter. (In the 1970s, it still permitted effective communication on the most crowded amateur bands.) Consumption was 6.5 amperes at transmission, key depressed, and 4.7 amperes during reception.

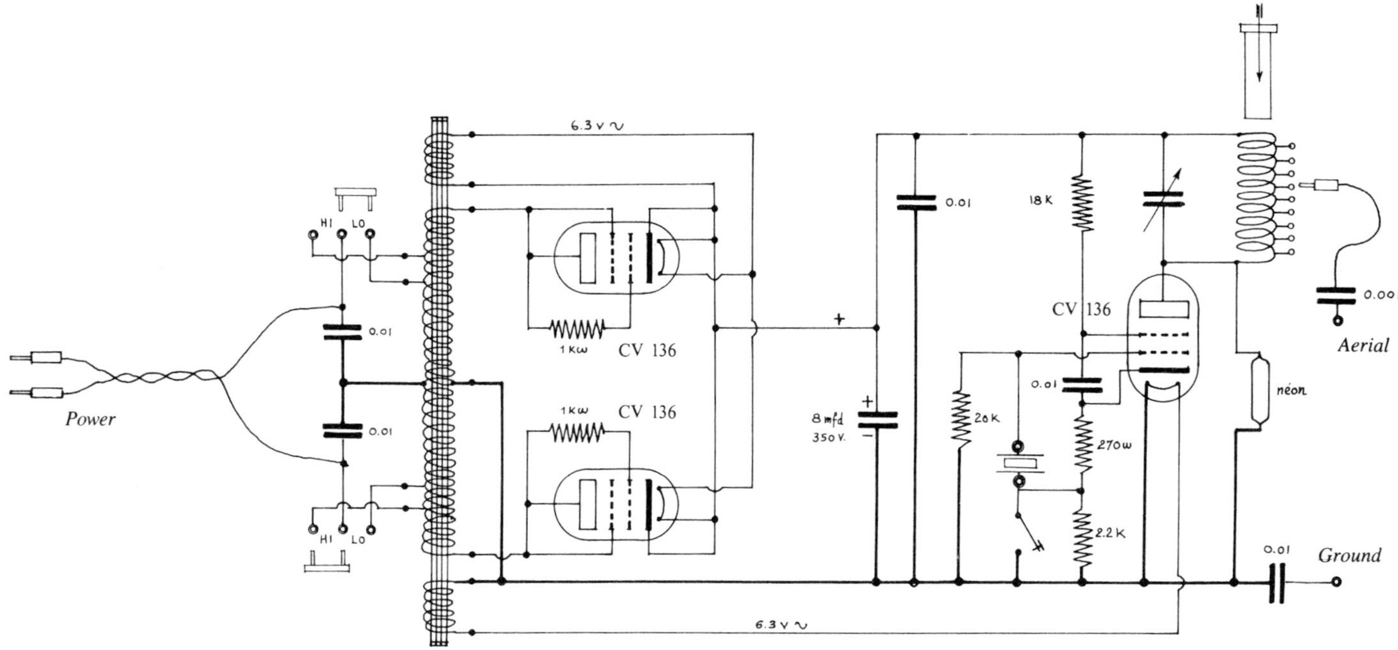


## Circuit Diagram of the 51/1 Miniature Transmitter

The transmitter was equipped with 3 miniature CV136 tubes and consisted of 1 crystal oscillator stage and built-in power supply consisting of an autotransformer and 2 tubes providing full rectification.

The technical design is especially brilliant. The number of components was reduced to the absolute minimum. Power consumption was 17 to 19 watts with the key depressed. The power radiated by the aerial was 3 to 4 watts.

An emergency transmitter for operating under extremely unsafe conditions, this set could not be asked to perform like a B2 or even an A Mark III. Nevertheless, its signal was of sufficient quality so that it did not cause too many problems for the operators in the large British home stations. However, the users of this set were advised not to use the crystal on its harmonic frequencies and to make themselves tap out their messages very distinctly while not transmitting too rapidly.

Its exceptional weight and size advantages were more than enough compensation for its weaknesses. Finally, if the oscillator tube happened to give up the ghost, the radio operator could possibly replace it with a rectifier tube. Only half-rectification was then attained, the tone was of only mediocre quality, and the aerial power fell to two-thirds normal, or 2 to 3 watts, but the transmitter still fulfilled its mission.

Today, the use of transistors would undoubtedly permit a further decrease in size and weight, although even with the 51/1, slender and delicate fingers were required for operating. However, the biggest advantage would be that the transmitter could be operated with a small 12-volt battery since the need to have a monotension power supply limited the 51/1's use.

FIG. 5

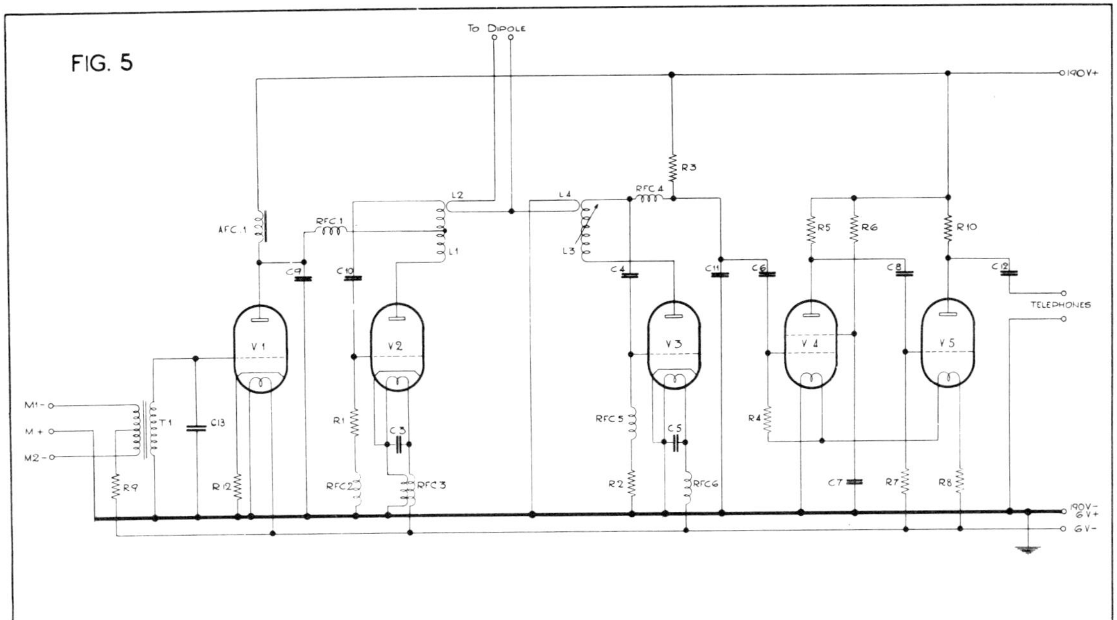


| | | | | | | | | | | | |
|---|---|---|---|---|---|---|---|---|---|---|---|
| L1 | Tx Osc. Coil | R4 | 2·2 MΩ ± 20% | | | C9 | 50pF ± 20% Ceramic | RFC2 | Tx Grid Choke | V3 | RL 18 (NR 88) |
| L2 | Aerial Coupling | R5 | 470 KΩ ± 10% | | | C10 | 10pF ± 20% Ceramic | RFC3 | Tx Filament Choke | V4 | XV5 2v Hivac |
| L4 | Coil Assembly | R6 | 1 MΩ ± 20% | | | C11 | ·005 μF ± 20% Moulded | RFC4 | Rx HT Choke | V5 | XP 2v Hivac |
| L3 | Rx Osc. Coil | R7 | 2·2 MΩ ± 20% | C3 | 50pF ± 20% Ceramic | C12 | ·01 μF ± 20% Tubular | RFC5 | Rx Grid Choke | | |
| | | R8 | 22 Ω ± 20% | C4 | 25pF ± 20% Ceramic | C13 | 0·1 μF ± 20% Tubular | RFC6 | Rx Filament Choke | T1 | Mic. Transformer |
| | | R9 | 15 Ω ± 20% | C5 | 50pF ± 20% Ceramic | | | | | AFC1 | Tx Mod Choke. |
| R1 | 8,200 Ω ± 10% | R10 | 150 KΩ ± 20% | C6 | ·0005 μF ± 20% Moulded | | | | | | |
| R2 | 150 KΩ ± 20% | R11 | | C7 | ·01 μF ± 20% Tubular | | | V1 | RL 18 M (NR 88) | | |
| R3 | 10 KΩ ± 20% | R12 | 1000 Ω ± 20% | C8 | ·0005 μF ± 20% Moulded | RFC1 | Tx. HT Choke. | V2 | RL 18 (NR 88) | | |

| Description | Drawn | Approved | Scale | Date | Drawing No |
|---|---|---|---|---|---|
| Model 13/IV S-Phone Circuit Diagram | ACG | | — | 23·12·43 | CD 2047 |

# Circuit Diagram of the S-Phone (Ground Equipment) 13 Mark IV (1944)

This was the apparatus carried by the agent in the field. The transmitter (circuit at left) was equipped with two subminiature tubes: the RL 18 M modulator and the RL 18 auto-oscillator.

The receiver (circuit at right) was a superregenerative receiver with subminiature tubes: the RL 18 detector, the XV5 audio-frequency amplifier and an XP second audio-frequency amplifier. The tubes had their electrodes sealed inside a tiny glass bulb and weighed a few grams.

The earlier models (13 Mark II and Mark III) had only 4 tubes, and the low modulation rating (20 percent) greatly limited the radiotelephone's transmitting range. The 1944 Mark IV, with its special circuitry, had a modulation rating of 40 percent. Although this did not change the range of the homing beacon, it increased radiotelephone range 5 to 7 miles more than that of previous models.

The receiver on the plane, not facing the same size and weight requirements, was a superheterodyne receiver of great sensitivity with 13 tubes and an electronic antenna switch.

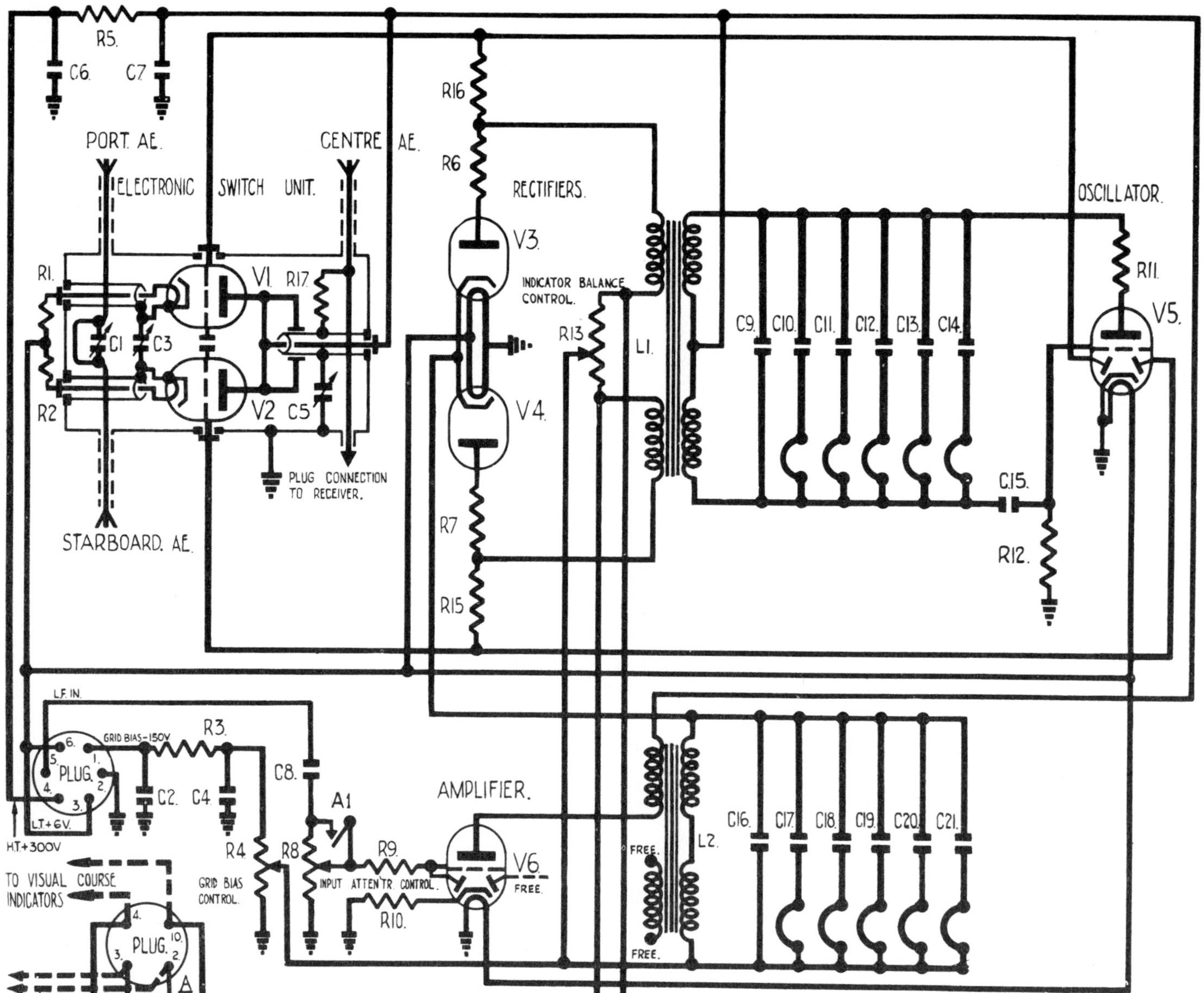


FIG.5. AERIAL SWITCHING PANEL.

## Circuit Diagram of the S-Phone (Aerial Switching Panel)

This is the circuit diagram of the electronic switch that separated the homing device operation from the radiotelephone operation on board the supply plane.

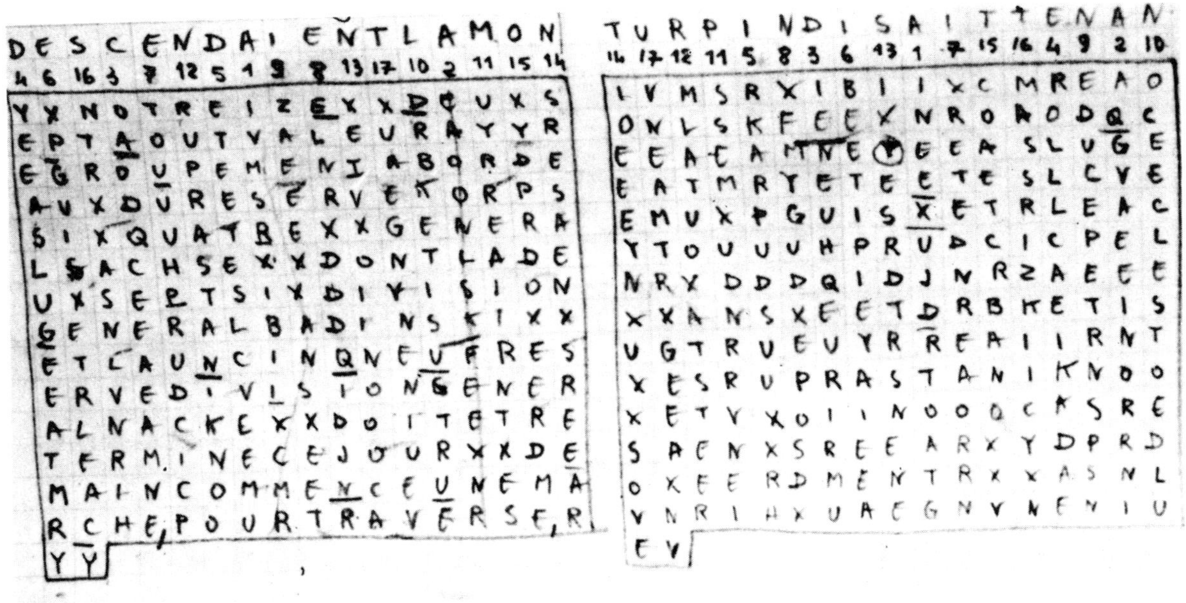


## Worksheet for the Encipherment of a Message in the Double Transposition System

Origin: French KLÉBER intelligence unit, southwest region (where the system was still in use).
Destination: London.

The text of telegram 13, dated August 27, 1944, read in French as follows: "No. 13, August 27, value level [of information]: A. Regroupment of Reserve Corps 64, led by General Sachse (General Badinski led the 276th Division and General Nacke, the 159th Reserve Division), should be completed today. A march begins tomorrow to cross . . ."

The complete text had been broken up to cut down on the length of transmission. The rest of the message comprised telegram 14 which was transmitted during another session.

The table on the left is the first transposition. At the top is a phrase from a book, which was used as the first key: DESCENDAIENT LA MON . . .

The text of the message is written horizontally.

Note the dummy letters, *yx*, placed at the beginning of the text as well as the procedure particular to the French services of separating sentences with doubled letters: *xx* or *yy*, instead of the word *stop* employed by the British services.

The right-hand table is the second transposition with the new key, TURPIN DISAIT TENAN . . . The columns of the first table are transcribed vertically extracted from top to bottom, in numerical order, and written horizontally into the second table.

The ciphertext to be transmitted will be taken out vertically by columns from top to bottom in the numerical order of the key: INEEX UJDRT OATGA QGVAE EINOR, etc. Code and verification groups should, of course, precede it.

The use of school notebooks with square-ruled paper simplified the cipherer's task considerably. Unfortunately, they were almost impossible to find, at least in Parisian stationery shops.

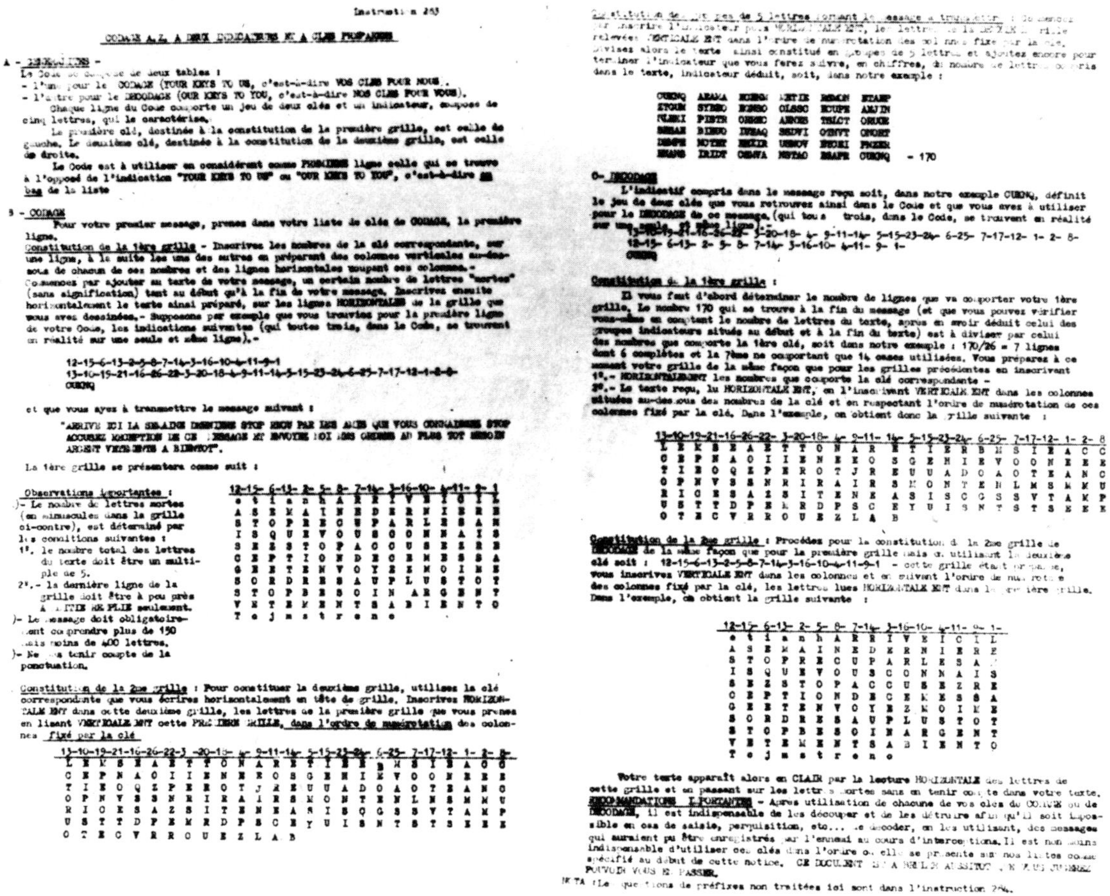

Instruction 283

CODAGE A.Z. A DEUX INDICATEURS ET A CLES PREPAREES

A - GENERALITES -

Le Code se compose de deux tables :
- l'une pour le CODAGE (YOUR KEYS TO US, c'est-à-dire VOS CLES POUR NOUS).
- l'autre pour le DECODAGE (OUR KEYS TO YOU, c'est-à-dire NOS CLES POUR VOUS).

Chaque ligne du Code comporte un jeu de deux clés et un indicateur, composé de cinq lettres, qui le caractérise.

La première clé, destinée à la constitution de la première grille, est celle de gauche. La deuxième clé, destinée à la constitution de la deuxième grille, est celle de droite.

Le Code est à utiliser en considérant comme PREMIERE ligne celle qui se trouve à l'opposé de l'indication "YOUR KEYS TO US" ou "OUR KEYS TO YOU", c'est-à-dire au bas de la liste

B - CODAGE

Pour votre premier message, prenez dans votre liste de clés de CODAGE, la première ligne.

Constitution de la 1ère grille - Inscrivez les nombres de la clé correspondante, sur une ligne, à la suite les uns des autres en préparant des colonnes verticales au-dessous de chacun de ces nombres et des lignes horizontales coupant ces colonnes.- Commencez par ajouter au texte de votre message, un certain nombre de lettres "mortes" (sans signification) tant au début qu'à la fin de votre message. Inscrivez ensuite horizontalement le texte ainsi préparé, sur les lignes HORIZONTALES de la grille que vous avez dessinées.- Supposons par exemple que vous trouviez pour la première ligne de votre Code, les indications suivantes (qui toutes trois, dans le Code, se trouvent en réalité sur une seule et même ligne).-

12-15-6-13-2-5-8-7-14-3-16-10-4-11-9-1
13-10-19-21-16-26-22-3-20-18-4-9-11-14-5-15-23-24-6-25-7-17-12-1-2-8-
CUBNQ

et que vous ayez à transmettre le message suivant :

"ARRIVE ICI LA SEMAINE DERNIERE STOP RECU PAR LES AMIS QUE VOUS CONNAISSEZ STOP ACCUSEZ RECEPTION DE CE MESSAGE ET ENVOYEZ MOI DES ORDRES AU PLUS TOT BESOIN ARGENT VETEMENTS A BIENTOT".

La 1ère grille se présentera comme suit :

Observations importantes :
1)- Le nombre de lettres mortes (en minuscules dans la grille ci-contre), est déterminé par les conditions suivantes :
1°. le nombre total des lettres du texte doit être un multiple de 5.
2°.- la dernière ligne de la grille doit être à peu près A MOITIE REMPLIE seulement.
2)- Le message doit obligatoirement comprendre plus de 150 mais moins de 400 lettres.
3)- Ne pas tenir compte de la ponctuation.

| 12 | 15 | 6 | 13 | 2 | 5 | 8 | 7 | 14 | 3 | 16 | 10 | 4 | 11 | 9 | 1 |
|---|---|---|---|---|---|---|---|---|---|---|---|---|---|---|---|
| e | t | i | a | n | h | A | R | R | I | V | E | I | C | I | L |
| A | S | E | M | A | I | N | E | D | E | R | N | I | E | R | E |
| S | T | O | P | R | E | C | U | P | A | R | L | E | S | A | N |
| I | S | Q | U | E | V | O | U | S | C | O | N | N | A | I | S |
| S | E | Z | S | T | O | P | A | C | C | U | S | E | Z | R | E |
| C | E | P | T | I | O | N | D | E | C | E | M | E | S | S | A |
| G | E | E | T | E | N | V | O | Y | E | Z | M | O | I | N | E |
| S | O | R | D | R | E | S | A | U | P | L | U | S | T | O | T |
| S | T | O | P | B | E | S | O | I | N | A | R | G | E | N | T |
| V | E | T | E | M | E | N | T | S | A | B | I | E | N | T | O |
| T | e | j | m | s | t | r | e | n | o | | | | | | |

Constitution de la 2me grille : Pour constituer la deuxième grille, utilisez la clé correspondante que vous écrirez horizontalement en tête de grille. Inscrivez HORIZONTALEMENT dans cette deuxième grille, les lettres de la première grille que vous prenez en lisant VERTICALEMENT cette PREMIERE GRILLE, dans l'ordre de numérotation des colonnes fixé par la clé

| 13 | 10 | 19 | 21 | 16 | 26 | 22 | 3 | 20 | 18 | 4 | 9 | 11 | 14 | 5 | 15 | 23 | 24 | 6 | 25 | 7 | 17 | 12 | 1 | 2 | 8 |
|---|---|---|---|---|---|---|---|---|---|---|---|---|---|---|---|---|---|---|---|---|---|---|---|---|---|
| L | E | K | S | E | A | E | T | T | O | N | A | R | E | T | I | E | R | B | M | S | I | E | A | C | C |
| C | E | P | N | A | O | I | I | E | N | E | E | O | S | G | E | H | I | E | V | O | O | N | E | E | E |
| T | I | E | O | Q | E | P | E | R | O | T | J | R | E | U | U | A | D | O | A | O | T | E | A | N | C |
| O | P | N | V | S | S | N | R | I | R | A | I | R | S | M | O | N | T | E | N | L | M | S | M | M | U |
| R | I | O | E | S | A | Z | S | I | T | E | N | E | A | S | I | S | C | G | S | S | V | T | A | M | P |
| U | S | T | T | D | P | E | M | R | D | P | S | C | E | Y | U | I | S | N | T | S | T | S | E | E | E |
| O | T | E | C | V | R | R | O | U | E | Z | L | A | B | | | | | | | | | | | | |

Constitution des groupes de 5 lettres formant le message à transmettre : Commencez par inscrire l'indicateur puis HORIZONTALEMENT, les lettres de la DEUXIEME grille relevées VERTICALEMENT dans l'ordre de numérotation des colonnes fixé par la clé. Divisez alors le texte ainsi constitué en groupes de 5 lettres et ajoutez encore pour terminer l'indicateur que vous ferez suivre, en chiffres, du nombre de lettres compris dans le texte, indicateur déduit, soit, dans notre exemple :

CUBNQ ARAMA EOEGE AETIE EMMON ETAMP
ETOUM SYEEO EOREO CIESC ECUPE AMJIN
NLEEI PIETR OEEEC AEECE TEICT ORUCE
EEEAN DIEEO IPEAQ SEOVI OTEVT ONOET
DEGPE MOTEE EEEIR UEEOV EEOEI PNZER
EEANS IRIDT OEMVA MSTAO EEAPE CUBNQ - 170

C- DECODAGE

L'indicatif compris dans le message reçu soit, dans notre exemple CUBNQ, définit le jeu de deux clés que vous retrouvez ainsi dans le Code et que vous avez à utiliser pour le DECODAGE de ce message.(qui tous trois, dans le Code, se trouvent en réalité sur une seule et même ligne).-

13-10-19-21-16-26-22- 3-20-18- 4- 9-11-14- 5-15-23-24- 6-25- 7-17-12- 1- 2- 8-
12-15- 6-13- 2- 5- 8- 7-14- 3-16-10- 4-11- 9- 1-
CUBNQ

Constitution de la 1ère grille :

Il vous faut d'abord déterminer le nombre de lignes que va comporter votre 1ère grille. Le nombre 170 qui se trouve à la fin du message (et que vous pouvez vérifier vous-même en comptant le nombre de lettres du texte, après en avoir déduit celui des groupes indicateurs situés au début et à la fin du texte) est à diviser par celui des nombres que comporte la 1ère clé, soit dans notre exemple : 170/26 = 7 lignes dont 6 complètes et la 7ème ne comportant que 14 cases utilisées. Vous préparez à ce moment votre grille de la même façon que pour les grilles précédentes en inscrivant
1°.- HORIZONTALEMENT les nombres que comporte la clé correspondante -
2°.- Le texte reçu, lu HORIZONTALEMENT, en l'inscrivant VERTICALEMENT dans les colonnes situées au-dessous des nombres de la clé et en respectant l'ordre de numérotation de ces colonnes fixé par la clé. Dans l'exemple, on obtient donc la grille suivante :

| 13 | 10 | 19 | 21 | 16 | 26 | 22 | 3 | 20 | 18 | 4 | 9 | 11 | 14 | 5 | 15 | 23 | 24 | 6 | 25 | 7 | 17 | 12 | 1 | 2 | 8 |
|---|---|---|---|---|---|---|---|---|---|---|---|---|---|---|---|---|---|---|---|---|---|---|---|---|---|
| L | E | K | S | E | A | E | T | T | O | N | A | R | E | T | I | E | R | B | M | S | I | E | A | C | C |
| C | E | P | N | A | O | I | I | E | N | E | E | O | S | G | E | H | I | E | V | O | O | N | E | E | E |
| T | I | E | O | Q | E | P | E | R | O | T | J | R | E | U | U | A | D | O | A | O | T | E | A | N | C |
| O | P | N | V | S | S | N | R | I | R | A | I | R | S | M | O | N | T | E | N | L | M | S | M | M | U |
| R | I | O | E | S | A | Z | S | I | T | E | N | E | A | S | I | S | C | G | S | S | V | T | A | M | P |
| U | S | T | T | D | P | E | M | R | D | P | S | C | E | Y | U | I | S | N | T | S | T | S | E | E | E |
| O | T | E | C | V | R | R | O | U | E | Z | L | A | B | | | | | | | | | | | | |

Constitution de la 2me grille : Procédez pour la constitution de la 2me grille de DECODAGE de la même façon que pour la première grille mais en utilisant la deuxième clé soit : 12-15-6-13-2-5-8-7-14-3-16-10-4-11-9-1 - cette grille étant préparée, vous inscrivez VERTICALEMENT dans les colonnes et en suivant l'ordre de numérotation des colonnes fixé par la clé, les lettres lues HORIZONTALEMENT dans la première grille. Dans l'exemple, on obtient la grille suivante :

| 12 | 15 | 6 | 13 | 2 | 5 | 8 | 7 | 14 | 3 | 16 | 10 | 4 | 11 | 9 | 1 |
|---|---|---|---|---|---|---|---|---|---|---|---|---|---|---|---|
| e | t | i | a | n | h | A | R | R | I | V | E | I | C | I | L |
| A | S | E | M | A | I | N | E | D | E | R | N | I | E | R | E |
| S | T | O | P | R | E | C | U | P | A | R | L | E | S | A | N |
| I | S | Q | U | E | V | O | U | S | C | O | N | N | A | I | S |
| S | E | Z | S | T | O | P | A | C | C | U | S | E | Z | R | E |
| C | E | P | T | I | O | N | D | E | C | E | M | E | S | S | A |
| G | E | E | T | E | N | V | O | Y | E | Z | M | O | I | N | E |
| S | O | R | D | R | E | S | A | U | P | L | U | S | T | O | T |
| S | T | O | P | B | E | S | O | I | N | A | R | G | E | N | T |
| V | E | T | E | M | E | N | T | S | A | B | I | E | N | T | O |
| T | e | j | m | s | t | r | e | n | o | | | | | | |

Votre texte apparaît alors en CLAIR par la lecture HORIZONTALE des lettres de cette grille et en passant sur les lettres mortes sans en tenir compte dans votre texte.

RECOMMANDATIONS IMPORTANTES - Après utilisation de chacune de vos clés du CODAGE ou de DECODAGE, il est indispensable de les découper et de les détruire afin qu'il soit impossible en cas de saisie, perquisition, etc... de décoder, en les utilisant, des messages qui auraient pu être enregistrés par l'ennemi au cours d'interceptions. Il est non moins indispensable d'utiliser ces clés dans l'ordre où elles se présentent sur nos listes comme spécifié au début de cette notice. CE DOCUMENT SERA BRULE AUSSITOT QUE VOUS JUGEREZ POUVOIR VOUS EN PASSER.

NOTA : Les questions de préfixes non traitées ici sont dans l'instruction 284.

## Instructions for the A–Z System

This is the British secret service's microfilm drawn up in French for B.C.R.A. agents.

It explains the double transposition system at its last stage of evolution, in which the numerical keys are prepared in advance and printed on a silk handkerchief.

The last line specifies that "this document is to be burned as soon as you decide you can get along without it."

One will compare the complexity of this system, which is not very safe, to the absolute simplicity of the undecipherable "one-time pad" summarized on the following page.

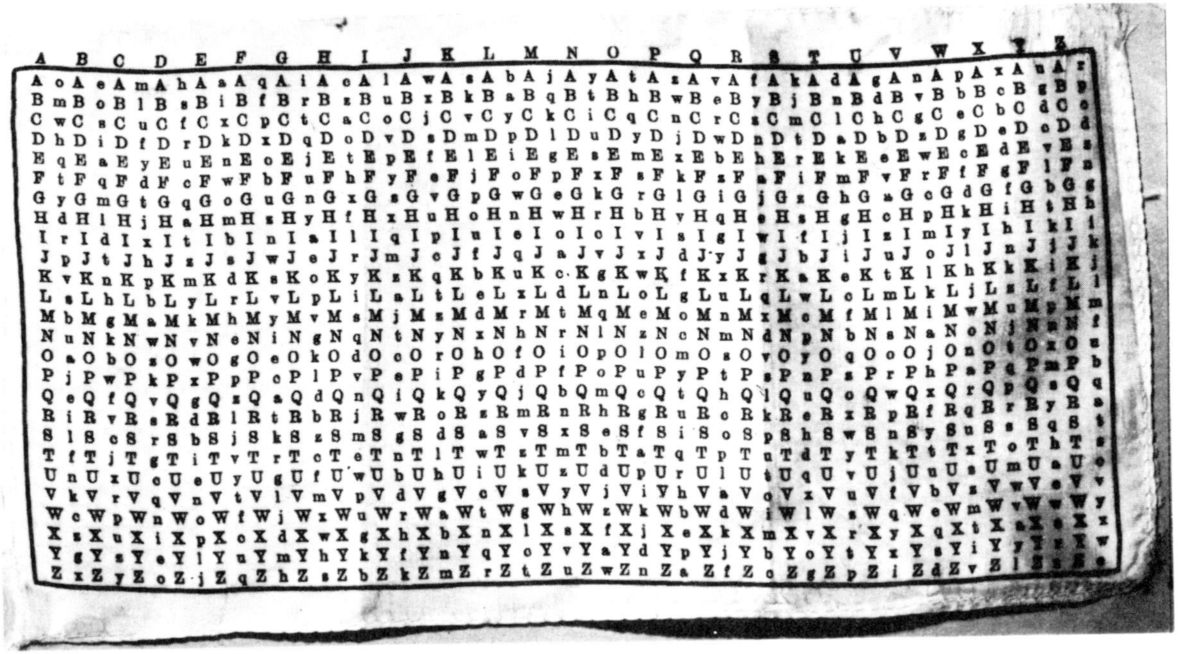


## S.O.E. F Section One-Time Pad System (1943–1945)

The grid is printed on silk (2/3 scale): the silk handkerchief is uncreasable. It can be rolled up into a very slender cylinder and inserted into a razor or brush handle, or any other ordinary article specially fixed for this purpose.

The "one-time pad" sheet (slightly reduced in size): this is half of a sheet used for ciphering messages transmitted from the outstation to the home station.

A complete set of keys consisted of 10 sheets numbered *1* through *5* for enciphering and *6* through *10* for deciphering. These sheets were supposed to be laid out flat for hiding, separated from the handkerchief, and kept within the immediate reach of the agent's hand so that they could be destroyed in the event of a search. The special treatment of the paper permitted its instant destruction by water or flame.

Keeping these incoherent series of letters in one's possession was obviously dangerous, but the immense advantage resulting from the system's unbreakability largely compensated for this danger which was, after all, minimal.

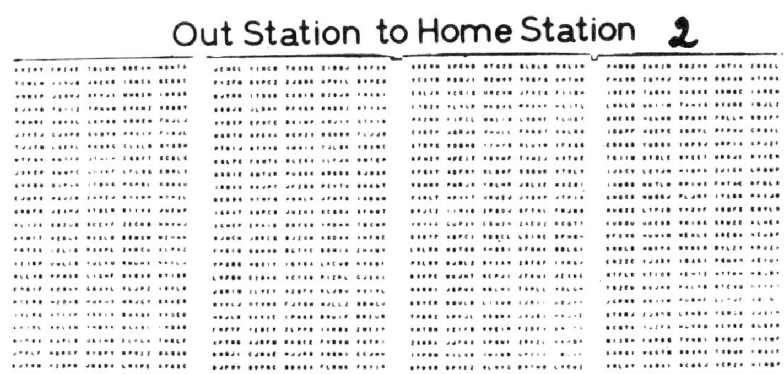

# Parachute Drops, April 1943 to September 1944

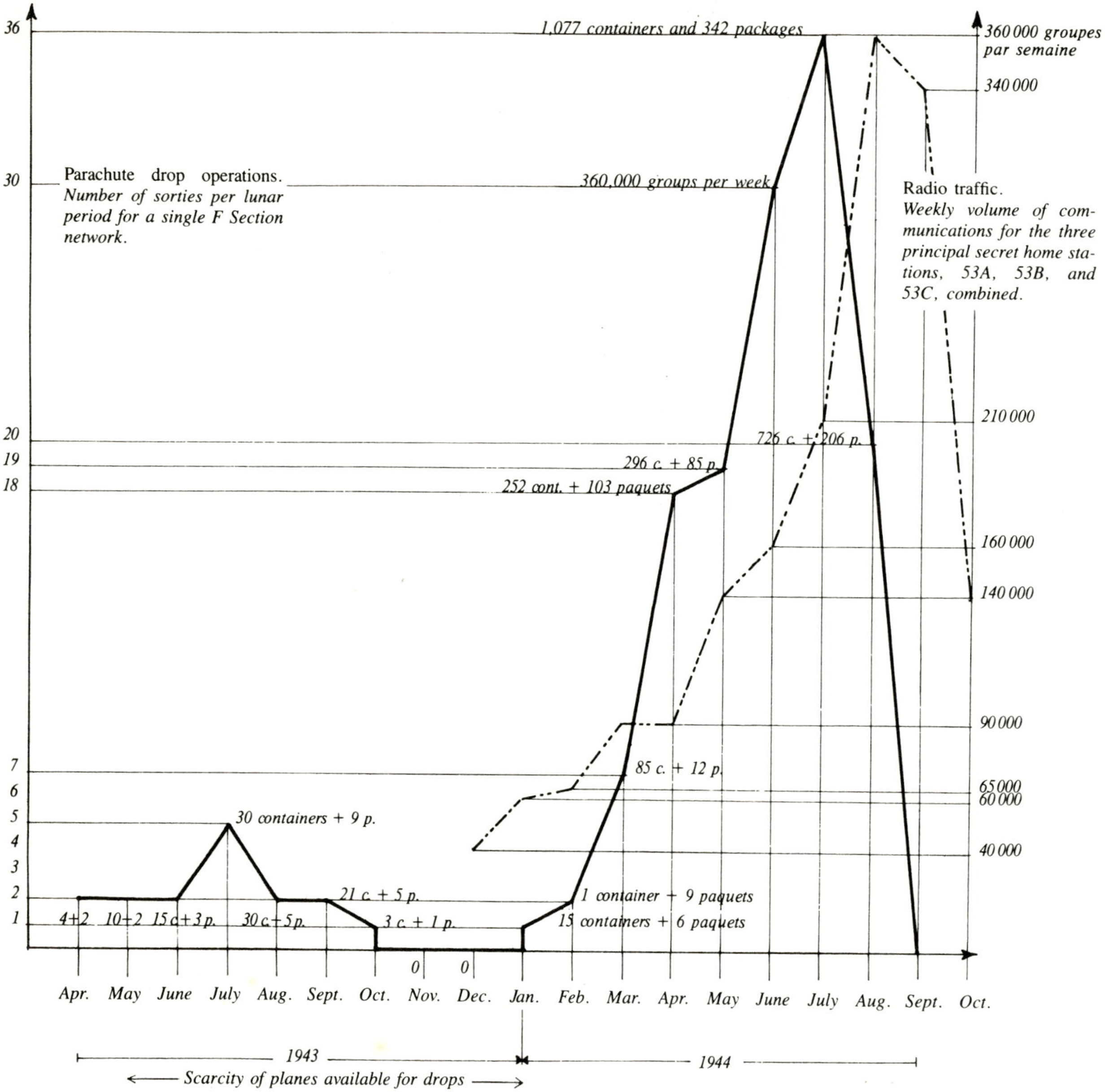


The solid curve shows the number of planes used as well as the number of containers and packages dropped for the armament of important circuits in the center of France in terms of lunar periods.

The curve only represents a single network (one of the most important). The curves which might be set down for other networks have the same shape and differ only in the amount of equipment delivered.

The dotted line shows the total volume of radio communication which flowed between the home stations and the clandestine outstations between December 1943 and October 1944. Note the perfect parallelism of the 2 curves. Radio communications peak one month after arms delivery peaks, which corresponds to the effective use of the arms on the average of one month after they are parachuted.

Note the intensity of radio communications. It must be specified that these are actual radio communications. It is likely that phantom operators were also transmitting dummy traffic to mislead the Germans about the actual number of orders and about command post locations. The same happened with the personal messages broadcast by the B.B.C. A variable but fairly large number of meaningless messages were transmitted every day to maintain volume fairly constant and to camouflage the significance of an increase in communications.



April May June July August September October November December January February March April May June July August

JEDBURGH TEAMS IN FRANCE JUNE - SEPTEMBER 1944.

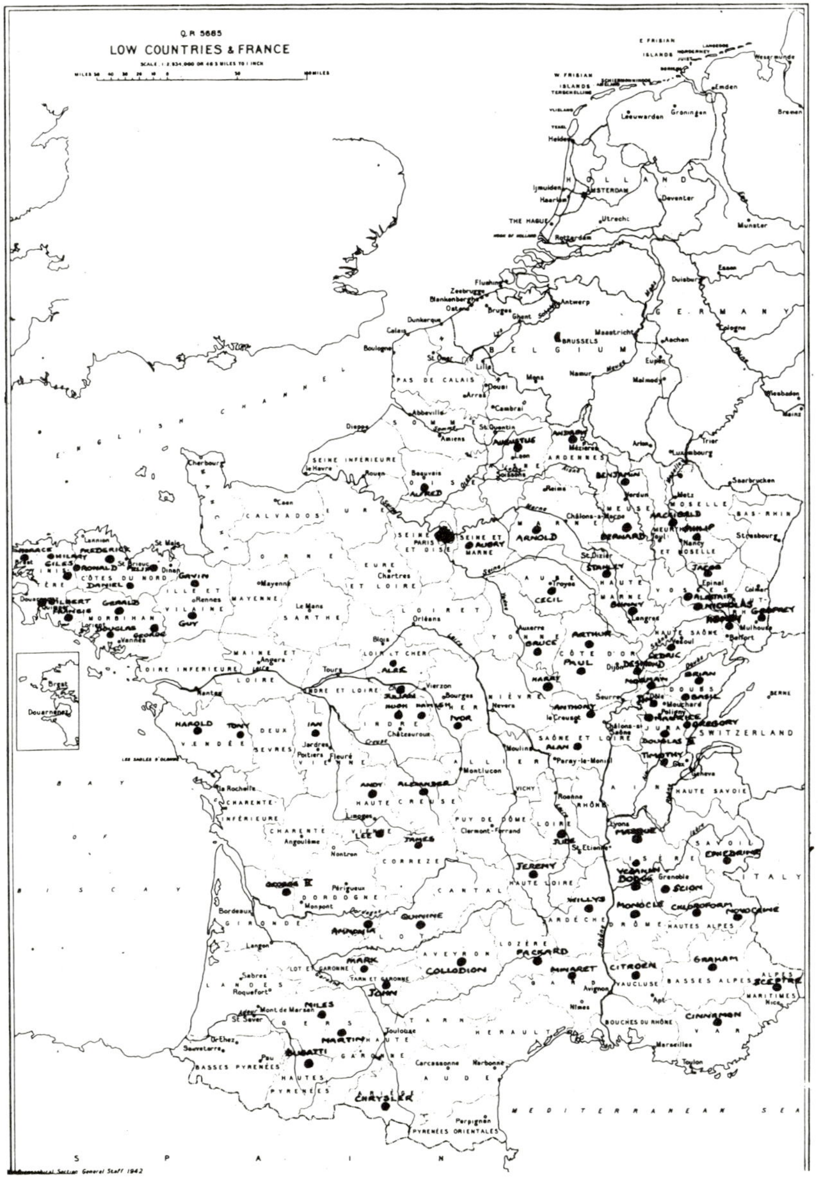


# The "Jedburgh" Teams

On D-Day, S.O.E. parachute drops permitted the outfitting of a substantial number of Maquis and several more or less isolated groups, whose number varied between a few dozen to several thousand. The problem that then posed itself was to ensure the organization and cohesion of this multitude in preparation for real operations of war, first of all, against the German reinforcement troops advancing toward the Normandy front, and afterward, when the time came, for delaying the enemy along his axes of retreat, for preparing for Allied troops' penetration, and in a more general way, for all clean-up missions on the home fronts.

Therefore, it was decided to drop missions of advice and liaison as reinforcement. These were called "Jedburgh" missions and were composed each of 3 men: an Englishman, an American, and a Frenchman, one of whom was a radio sergeant. About 100 American O.S.S. officers, specially trained in S.O.E. classes, formed part of these teams.

Between June and September 1944, 86 Jedburgh teams saw action at the key points marked on the map. The members of these teams were all in uniform; open war had broken out. On July 1, 1944, the F and RF sections of the S.O.E. merged with the B.C.R.A. This merger constituted the general staff (état-major) of the French Forces of the Interior (E.M.F.F.I.), commanded by General Koenig.

Besides the Jedburgh missions, about 20 "interallied missions" performed special assignments for the Allied troops. There were as many Franco-British teams as there were purely French teams.

The rare photographs taken of these missions, notably the "Aloes" mission, parachuted into Brittany in August 1944, and the "Poivre" mission, dropped outside of Laon in October 1944, show the use of 3 Mark II radio sets and Eurêka homing devices. The men are wearing English battle dress, often without rank badges, and a forage cap or beret. Each is armed with a Colt pistol and a U.S. M-1 carbine.

G3ZWH D.M. HILL, 16 HOLLOW LANE, SNODLAND, ME6 5LS.